THE COMPLETE GUIDE TO BUILDING WITH ROCKS & STONE

Stonework Projects and Techniques Explained Simply

REVISED 2nd Edition

BY BRENDA FLYNN

The Complete Guide to Building with Rocks & Stone: Stonework Projects and Techniques Explained Simply REVISED 2nd Edition

1405 SW 6th Ave. • Ocala, Florida 34471 • 800-814-1132 • 352-622-1875–Fax
Website: www.atlantic-pub.com • E-mail: sales@atlantic-pub.com
SAN Number: 268-1250

Library of Congress Cataloging-in-Publication Data

Flynn, Brenda, 1955-
The complete guide to building with rocks & stone : stonework projects and techniques explained simply / By Brenda Flynn. -- REVISED 2nd Edition.
pages cm
Includes bibliographical references and index.
ISBN 978-1-62023-015-2 (alk. paper) -- ISBN 1-62023-015-1 (alk. paper) 1. Stonemasonry--Amateurs' manuals. I. Title.
TH5411.F58 2015
693'.1--dc23

2015024567

2ND EDITION EDITOR: Melissa Shortman • mfigueroa@atlantic-pub.com
ART DIRECTOR: Meg Buchner • megadesn@mchsi.com

Printed in the United States

Reduce. Reuse. RECYCLE.

A decade ago, Atlantic Publishing signed the Green Press Initiative. These guidelines promote environmentally friendly practices, such as using recycled stock and vegetable-based inks, avoiding waste, choosing energy-efficient resources, and promoting a no-pulping policy. We now use 100-percent recycled stock on all our books. The results: in one year, switching to post-consumer recycled stock saved 24 mature trees, 5,000 gallons of water, the equivalent of the total energy used for one home in a year, and the equivalent of the greenhouse gases from one car driven for a year.

Over the years, we have adopted a number of dogs from rescues and shelters. First there was Bear and after he passed, Ginger and Scout. Now, we have Kira, another rescue. They have brought immense joy and love not just into our lives, but into the lives of all who met them.

We want you to know a portion of the profits of this book will be donated in Bear, Ginger and Scout's memory to local animal shelters, parks, conservation organizations, and other individuals and nonprofit organizations in need of assistance.

– Douglas & Sherri Brown,
President & Vice-President of Atlantic Publishing

Table of Contents

Chapter 2: Working with Stone Tools 51

Chapter 3: Working with Dry Stone and Mortar 95

Chapter 4: Patios and Outdoor Cooking Pits 107

Chapter 5: Enhancing Your Landscape 129

Chapter 6: Building Stone Walls 153

Chapter 7: Creating Rock Gardens 183

Chapter 8: Using Stone to Build Bridges and Dams 207

Chapter 9: Fireplaces, Entry Gates, and Mailboxes 223

Chapter 10: Restoring Existing Stone Structures 247

Chapter 11: Caring for Stone 255

Author's Commentary

"A man who works with his hands is a laborer; a man who works with his hands and his brain is a craftsman; but a man who works with his hands and his brain and his heart is an artist."
– Louis Nizer

As a child, I remember watching my father work with concrete and stone, pouring slabs for the foundation of our home, which he built from the ground up. I was fascinated by how he just knew how much cement to put in the huge, metal mixing trough, and when he added water, he used the garden hose, never measuring, working a hoe in a back-and-forth fashion until the mixture was just right. He fashioned stones for the patio from concrete, poured the foundation for our carport and sidewalks, and created steps. He built barbecue pits, stone walls, and rock gardens. He put stone and brick veneers on our A-frame house.

I took this knowledge and subsequently built brick planters and stone walls for my own home later in life and learned the trade of setting tile, creating entire floors for my patios, bathrooms, and later, my art gallery. I do mosaics and love to work with stone in my art.

I always collect rocks and stone from all my travels, from the granite country of Maine to the Irish Sea on the south of Ireland where miles of marvelously rounded stones exist. I have rocks from the Appalachian Mountains holding books in place on my bookshelf and have built miniature cairns as paperweights on my desk. There is never an instance I do not get dazzled by a particularly unusually shaped stone and pick it up to put in my pocket.

INTRODUCTION: The Beauty and Strength of Stone

Because of its superior strength, beauty, and longevity, stone has provided a readily available, natural building material for thousands of years. Stone structures appeal to us psychologically because they represent security and provide a sense of protection. Stone is also one of the oldest construction materials known to man. Excavations of stone circles in Great Britain and Eastern Europe are believed to be rudimentary living structures that date back to 12000 B.C. Early stonemasons in Egypt built the pyramids from hand-quarried rock beginning in about 2600 B.C., applying techniques of **dry stone** masonry — not using an adhesive cement called **mortar** — that people still use today.

Rock is simple, elegant, and, in the long-term, very cost effective to use for almost all building applications. Asphalt roads crumble and pit, forming holes that enlarge and cause potholes, which require constant maintenance. But, roads made from stone and brick, such as the famous cobblestone streets of Bath, England built during the Roman inhabitation before A.D. 407, have withstood countless travelers in many forms of vehicles, including horses, carriages, and the modern-day automobile. British medieval castles and French arched bridges made from stone have required very little or no repair since people first built them in the early 1500s. Stone is weatherproof, vermin proof, fireproof, insect proof, and resists water damage. When damage does occur because of natural aging or an act of man, people can easily repair the stone without having to completely rebuild the structure.

On the negative side, stone is heavy, cumbersome to move, and handling it can result in skinned hands, bruises, and if you are not careful, broken bones. It is also difficult to shape to fit into a certain nook or cranny. But, once you learn how to use tools to shape stone properly, it can all come together magnificently. This book will explain those tools and their uses, enabling you to complete your projects safely and efficiently.

Stone also has another undesirable quality: It is a poor source of insulation. Although stone houses have the reputation of remaining cool in the summer, which provides a comfortable living situation if you do not have central air conditioning, it does not retain heat for a long time. Heat flows through stone, and depending on the climate, water will condense on both inside and outside surfaces if not properly insulated. Therefore, if you build a stone structure for habitation or food storage, you should include some form of insulation, such as rolled fiberglass panels or blown insulation in between the inside and outside wall to prevent the growth of molds and mildew on the interior walls. Building something with stone also takes considerable time, and if you expect a quick afternoon project, do not use this medium, much less build an entire home out of it. The monetary and historical value of a structure built with stone will greatly increase because of its permanence and ease of maintenance.

Having read this, you are probably wondering if you can work with stone, especially if you have never mixed mortar, formed a simple wall, or you weigh less than 150 pounds. The answer is yes. Everyone has to start somewhere, and this book will help you build a project with stone and understand the nature of stone and where it comes from. A person who is not a carpenter can build with wood, a person who is not a potter can make clay pottery, and you, who are presumably not a stonemason, can build with stone. This book contains chapters outlining how to build a fireplace or fire pit; how to build benches and steps for your garden; how to build a dry stone wall for accenting your property and keeping animals in or out; and building rock and water gardens to complement your landscape. This book also incorporates outlines to take on larger feats, such as building stone dams and bridges. Whatever your project, whether big or small, it starts with the first step. Learning the tools of the trade and how you can use them to build your projects is a good place to start. This book will help you determine what projects to work on, how to start them, and will provide you with all the most valuable information, such as where to buy or rent your tools, how to use them, steps to start and complete your project, and time-honored tips and tricks to make your project a success. Now, it is time to get started.

This is a historic estate on Cumberland Island. As you can see, the stonework has outlasted everything else on the estate. Photo courtesy of Douglas R. Brown.

CHAPTER 1: Common Types of Stone and Rock

Many people think of stone as large boulders that create cliffs, buttes, and mountaintop ranges. Many may also think of landscaping stone in pleasant, rounded shapes used in flower beds and front entryways. Rock and stone are considered the same thing, but it is accepted that rock is stone in its purest form — without any smoothing, cutting, or shaping. Stone is the term used when rock is smoothed, chipped, or textured for landscaping or building. When you are looking to build with stone, you will find it comes in many shapes, weights, textures, colors, and sizes. This chapter describes the types of stone primarily used for architectural purposes and the unique properties of each.

Sedimentary Rock

Sedimentary rock is one of the three main rock groups — along with igneous and metamorphic — and is formed when remains of other rocks and biological activities, such as leaf and animal matter decomposition, known as **diagenesis**, are compressed. Common types of sedimentary rocks include chalk, coquina, limestone, sandstone, and shale. Chalk is not used as building material because it is very soft, but it is quarried to use in the manufacturing of writing chalk and used along with kaolin, a type of clay, in cosmetics and ceramic pottery. Most limestone contains fossil imprints because of diagenesis, and often these imprints are integrally maintained because sedimentary rock is not formed under the high temperatures and pressure that igneous and metamorphic rock are formed under. Sedimentary rock is formed very slowly over time when elements like broken shells, which contain a high amount of calcium and lime; plant materials; sand; grit; and other organic materials, are washed into rivers, lakes, oceans, and craters and these elements experience continuous pressure that climactic changes in earth core temperature or the continuing weight of additional elemental materials can cause. Over thousands of years, the bottom layers of washed away elements become rock.

The Earth's layers consist of an upper layer of sedimentary rock, a middle layer of igneous rock, and an inner layer of metamorphic rock.

Sedimentary rock forms a crust over the igneous and metamorphic rock and is much softer and easier to build with. Limestone and sandstone are the easiest types of stone to cut and shape for most building projects. These two stones reflect light and are used in many countries, such as Egypt and Syria, for just that purpose. Different types of sandstone and limestone range from fine-grained and hard to coarse-grained and crumbly. Because these stones were formed in layers over millions of years, they have a horizontal grain, and when the stone weathers and breaks, it breaks into large, rectangular blocks. These blocks are perfect for building walls and other similar structures requiring uniformity of stone. The following are examples of sedimentary rock.

Sandstone

Sandstone is by far the mason's choice because of how easy it is to use. You can carve it, chip it, or chisel it into whatever shape you need it molded into. Sandstone is unique among sedimentary rock because it naturally takes on a myriad of colors, depending on the location of where it is quarried. **Quarrying** is the term used to define the actual digging and harvesting of rock and usually takes place in mountainous areas where entire mountains are cut apart using heavy construction machinery designed for moving and lifting heavy loads of rock.

This is a working stone quarry that has belt conveyors and various mining equipment.

Different minerals and oxides in many areas of the country determine the rock's coloration. In the Ozark Mountains of Arkansas, the stone is popular as a building veneer on concrete block because of its vast array of coloration. **Veneer** is defined as a thin sheet of stone or stones and rock that are cemented to block or, in some early cases, wood. On Prince Edward Island, Canada, there is a type of sandstone naturally brick red in color due to the high concentration of iron oxide in the soil from which the rocks formed. In the Appalachian Mountains of West Virginia, sandstone exists that is a dark chocolate color in its natural state, but when it is quarried and weathers, it turns a dull gray color. Most sandstone colors are some form of gray, brown, white, rose, and slate-blue, depending on the region it comes from and the mineral contents of that region. When sandstone weathers, as in **fieldstone** — naturally occurring chunks of rock found in pasturelands and fields as opposed to quarried stone, which is stone dug from the earth by mechanical means — it becomes softened around the edges, giving it a rounded appearance.

Sandstone commonly breaks in pieces with flat tops and bottoms because as a sedimentary rock, it forms in layers and breaks back in layers when broken or quarried. These naturally worn stones are popular to use when building garden pathways and fireplace hearths because of the lack of sharp edges and abundance of flat surface area. Quarried sandstone exposed to the elements will weather somewhat quickly and achieve the characteristic worn look in a relatively short time, sometimes in just a matter of months, depending on the season or climate. One of the most prized properties of sandstone is that it is **porous**, meaning the rock is not dense and contains air pockets that collect water and air and help lichen and mosses grow. Ivy will root into the sandstone itself on building walls and thrive quite nicely. For this aesthetically pleasing reason, many gardeners love using the stone in their outdoor landscaping. Most of the quaint British countryside homes have been built with this stone, which allows ivy and moss to grow on its surface, giving it that desired Old World feel. The Smithsonian Museum Arts and Industries

Building in Washington, D.C. was built from red sandstone quarried from Seneca, Maryland and is as well known for its trademark ivy-covered walls as it is for the contents of the museum.

The Smithsonian Castle and Information Center was built out of red sandstone quarried from Seneca, Maryland.

Close-up of sandstone quarried from Fire State Park in Nevada.

Sandstone blocks were used to build this wall at the historic St Francis Xaviers Church, Berrima in New South Wales, Australia.

Here sandstone rock is used to create a waterfall. A hidden pump recycles the water back into the pool. Photo courtesy of Douglas R. Brown.

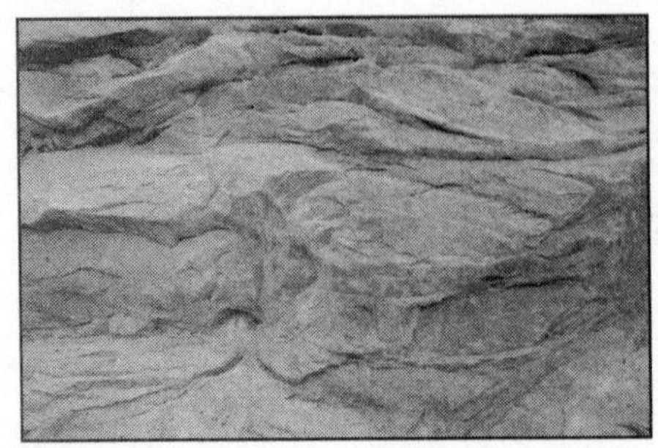

Close-up of a sandstone rock formation.

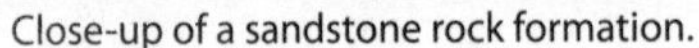

This wall was constructed out of different shades of sandstone.

Limestone

Limestone, a denser form of sandstone, was long the material of choice for commercial building in the United States and was even more popular than manufactured brick. Prior to the early 1900s when concrete block became the standard, limestone's hardness and durability made it used extensively. Brick manufactured in the early 1800s tended to be brittle and broke easily under pressure, while the quarried limestone maintained its integrity and strength.

Limestone is commonly used for carving grave monuments if granite is not used and, therefore, people associate the dark gray color of tombstones with this form of stone. But, some newly quarried limestone has a dull yellow color, although in many parts of the United States, primarily the Appalachian region, it is naturally a dark gray. A limestone found only in northern Kentucky has a light gray color with a blue hint, most likely because of a blue clay indigenous to the region.

Because of how and where limestone forms, imprints of things like leaves and prehistoric seashells are often found dotting the surfaces of the stone. Aged limestone appears almost corroded, although the pitted surfaces, which would seem to indicate weakness, belie the strength underneath. You can also use it in slabs as an attractive flooring material for both indoor and outdoor use, and, with a couple of coats of masonry sealer, it can last for centuries.

Limestone fieldstone is also the stone used most often in dry stonework. Many of these stones are evenly layered and naturally rectangular and brick-like in shape, allowing for easier fits when placed on top of other stone, as in dry stone walls.

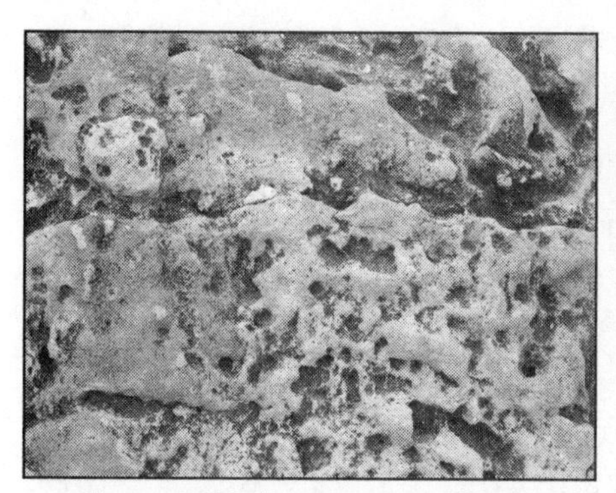
Close-up of Mediterranean coastal limestone.

Limestone rock was used to create a waterfall for a pool. Photo courtesy of Douglas R. Brown.

Close-ups of limestone rock. Photo courtesy of Douglas R. Brown.

Shale and slate

Shale and slate are soft-layered stones used for veneer on more durable manufactured stone, such as concrete block. These stones are not good as primary building stones because they tend to break apart under pressure and do not maintain their strength. Shale and slate are good to use as garden stepping stones and interior floor tiles in areas without high traffic flow. Harder shale and slate stones, quarried below the softer shale and slate and split into layered slabs, are commonly called **flagstones**. Flagstone is used for flooring on outdoor and indoor patios, high-traffic indoor flooring, and fireplace hearths. Flagstone comes naturally colored in many variations of blue, often referred to as bluestone; brown; red; mauve; and black. Also, a buff-colored flagstone is quarried primarily in New Mexico. Shale and slate is usually dark gray in color to an almost black.

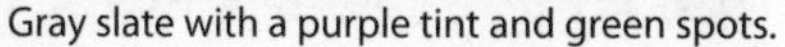

Gray slate with a purple tint and green spots.

Gray slate that has a rustic color and texture.

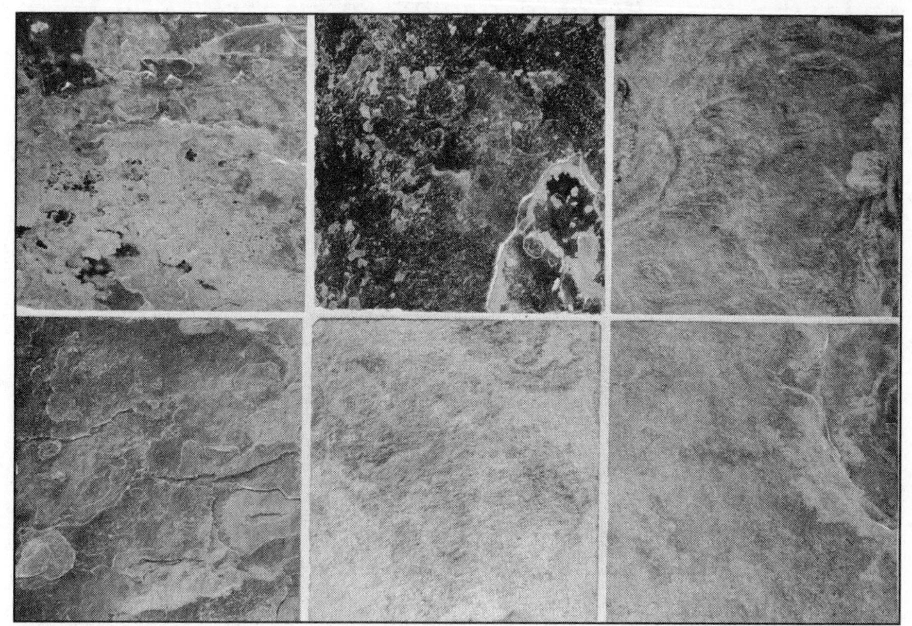

Different colored slate was used to create a tile floor in a house.

Coquina

Coquina is almost never mentioned as a building stone because of its regional availability, yet coastal states extensively use it. There are structures, such as the famous Castillo de San Marcos in St. Augustine, Florida, built entirely from this stone. This fort was built in the late 1600s, and although Florida is known for its harsh tropical climate, the fort still stands today as impenetrable as when the Spaniards used it to defend their holdings in Florida against the British and French. Coquina is mainly composed of the mineral calcite, coral, and phosphate in the form of tightly compacted seashells. It is found along the East Coast of Florida and may be found up to 20 miles inland from the coast in the subsurface, fewer than 4 feet under the earth. People can find coquina rock as far north as North Carolina and also on the South Island of New Zealand where coquina rock formations appear above the surface of the surrounding outcroppings of land along the coast. Several abandoned quarries in New Zealand contain coquina rock as well.

This is an aerial view of Castillo de San Marcos National Monument in St. Augustine, Florida. This fort was built out of coquina rock taken from Anastasia Island between 1672 and 1695.

People have quarried coquina and used it as a building stone in Florida for more than 400 years, and it has provided an exceptional building material for forts, specifically those built during periods of heavy cannon use from invading fleets. Because coquina is a soft rock material, cannon balls would sink into, rather than crush and shatter, fort walls.

When first quarried, coquina is extremely soft and somewhat damp, allowing for easy cutting and shaping. Coquina has to be left out in the weather to dry for approximately one to three years so the stone can harden because it is too soft for immediate use as a building stone. People have also used coquina as a source of paving material but now is only used as filler. They cut coquina into paving stones and laid it flat to make roads for vehicular and carriage traffic, but it was too soft for constant use because it developed breaks and worn ruts over just a few decades. In states where coquina is readily available, such as Florida and Alabama, people add it to asphalt to increase the quantity and to add strength to the mixture. Large rocks of coquina are sometimes used as landscape decoration.

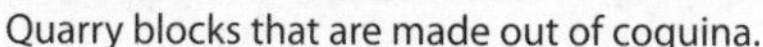
Quarry blocks that are made out of coquina.

Close-up shot of coquina stone.

Igneous Rock

Igneous rocks form when molten materials cool and solidify, usually below the Earth's surface. This is known as **intrusive rock**. The appearance of intrusive rock, such as granite, is mottled with large crystallizations of mica and glass. A slower cooling-down period inside the earth causes the large crystallizations. Alternately, some igneous rock forms above the surface when lava cools and hardens, creating new land surface. This is called **extrusive rock**. Solid volcanic lava, greenstone, and basalt are examples of extrusive rock, and crystallizations are much smaller. In some rocks, such as obsidian, you cannot detect them with the naked eye. Obsidian is as close to a naturally created glass as it gets in the rock world.

Igneous rocks are very hard and are used where texture and permanence are desired, such as in buildings and stone gate pillars. But, these are difficult to shape for building, especially by hand. Some basalt is so glass-like in structure it will actually slice skin. Beginning stonemasons should not use this form of rock in their projects.

Example of igneous rock found near a small village in the Atacama Desert. There is a volcanic plateau in Chile, South America west of the Andes Mountains that created this.

Aerial view of an igneous rock filled area in Maui, Hawaii.

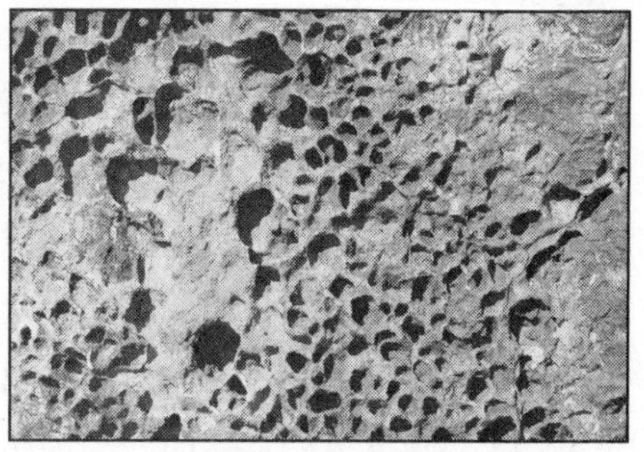

Close-up, detailed picture of igneous rock.

Granite

Granite is a coarse-grained, igneous rock, which makes it much harder and denser than sedimentary rock. It is difficult to work with because of its density. Recently, people started to associate granite with its use as countertops and flooring in modern homes, but in the late 1800s and early 1900s, it was used for impressive public buildings, such as museums, courthouses, and hospitals. Granite's natural strength and durability make this stone harder to work with because it has a variable texture that tends to break into less-desired shapes. To cut granite, a stonemason must use an air-powered cutting tool, and the dust generated from this procedure can be very hard on the lungs. With the advent of more complex machinery, such as massive pneumatic and hydraulic drills, you can now achieve precision cuts and thinner blocks of this stone without as much dust or accidental breakage.

The most common color for granite is a dark, speckled gray, but because of its high content of **feldspar** and **quartz**, natural minerals that affect the coloration of stone, it also takes on dark blue, greenish, and pink tones, both individually and within the same vein of rock. When granite breaks because of climactic changes or forces of nature, it breaks into huge stones weighing more than 500 pounds and up to a ton; therefore, most granite is now commercially quarried. Early pioneers and settlers have already picked up the smaller, more accessible fieldstone granite, smaller chunks of granite quarried commercially or naturally, and used it in building structures, such as springhouses — where dairy products and eggs were stored to remain cool — and root cellars — where many farmers kept their root crops and seeds for the following planting season. Many of the granite stones used in old foundations, chimneys, and basements are now recycled when the original structures are torn down or rebuilt. As long as you can move the stones, you can reuse the stones.

Kitchen that has granite countertops and back splashes.

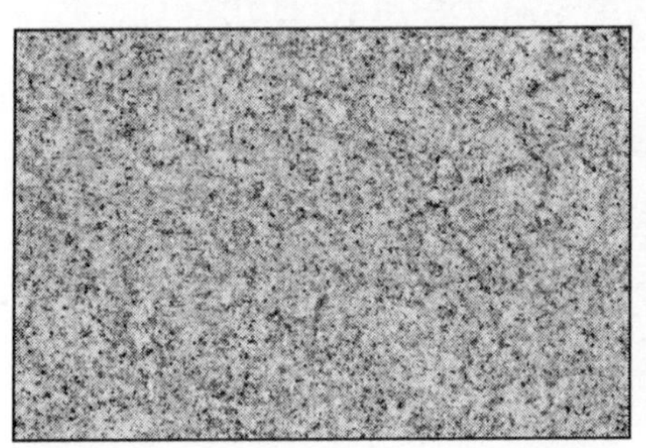

Brown and orange granite.

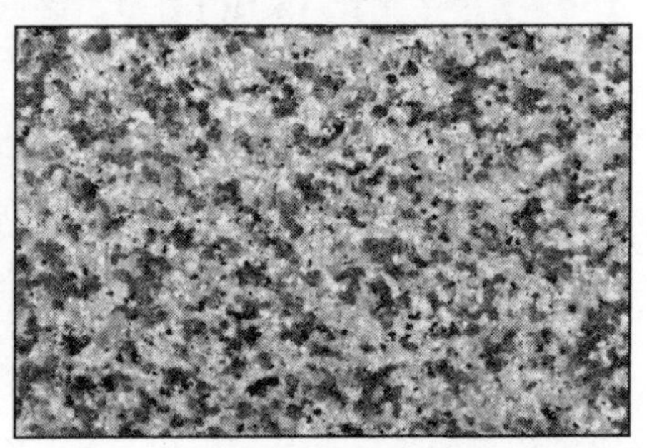

Close-up of the detail in granite.

Polished gray granite.

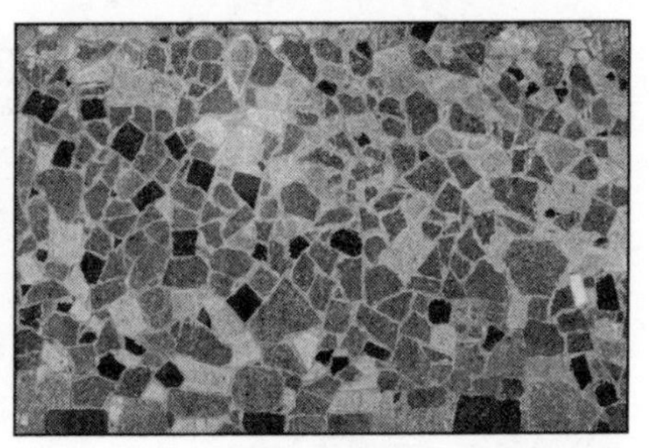

Panel of a decorative wall created from different colors of granite.

Basalt

Basalt is another commonly used stone classified as an extrusive stone and has the same properties of granite but without the grain. Basalt is much like obsidian in that it forms from rapidly cooled volcanic eruptions. Because of the rapid cooling process, basalt does not form interior strata but instead has a glass-like structure. It is an igneous rock and is very difficult to work with because it is so hard and is not easy to carve. But, it does possess an admirable density, with very little, if any porosity, and has a rich, black color. It is associated with many Japanese Zen gardens where large, carved basalt is often used as a meditation point. Crushed basalt is used as an aggregate in asphalt and is used by railways as ballast. Ballast is any heavy material, such as rock and stone, used to stabilize a ship. It is also used as coarse gravel laid to form a bed for streets and railroads. Basalt also forms most of the underlayment of the Earth's surface.

Close-up of the texture of basalt.

Years of eroding powers from the river created this basalt rock canyon.

Turret of an ancient fortress in Pico Island, Azores. The turret was built out of basalt.

Rock gate at the cliffs formed by volcanic basalt columns in Arnarstapi, Snaefellsnes Peninsula, Iceland.

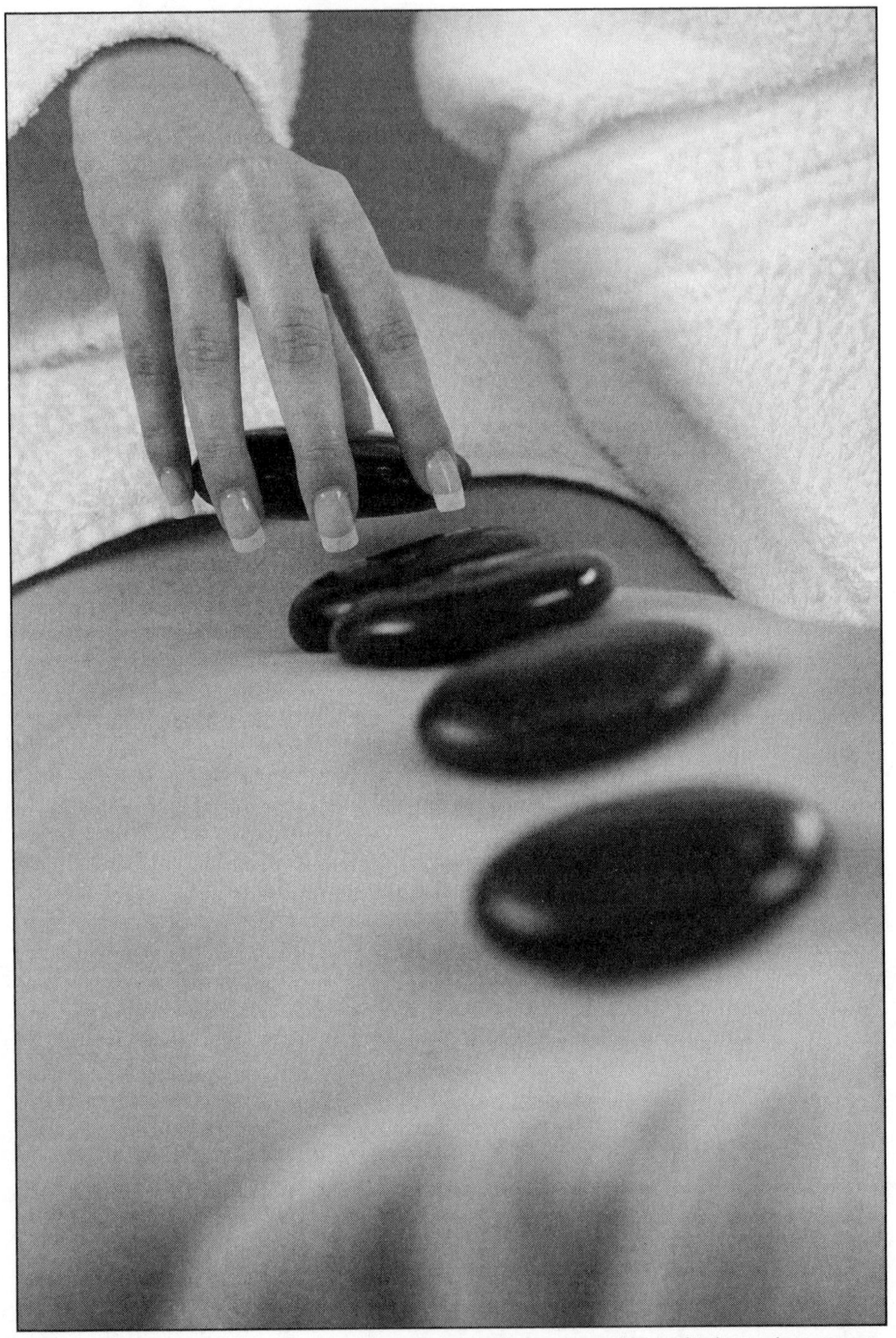

Basalt rocks are commonly used in hot rock treatments.

Greenstone

Greenstone is a type of basalt found in regions of the southeast United States, particularly the Blue Ridge Mountains. It merits mentioning because of its unique properties, not only in building but also because it adds the element of sound. When struck, most stones emit a dull noise. But, because it was formerly prehistoric sea lava and it contains a high concentration of metallic oxides, greenstone possesses the unique ability to emit a ringing sound, like a bell, when struck. It forms in long, greenish-blue shards and is very hard to carve or cut. It tends to break in thin shanks and can only be used as it is found. Being an igneous rock, it is very hard and durable, which makes it desirable, particularly for accent or veneer over concrete block or brick. Quarries usually blast out the stone and use it as landscaping accent, such as a rock placed in the middle of a sitting garden or as the base for a fountain or birdbath. It is also used for making **riprap**, which is irregular stone used for fill or to hold against erosion. People have made fireplaces and other indoor accents completely with greenstone found in fields, but it is a difficult stone to work with because it is so difficult to carve or cut.

Two spheres that were carved and shaped from a greenstone rock.

Geographic Locations of Natural Stone throughout the United States

Although you can find many types of stone, particularly sandstone and limestone, in almost all of the United States, there are certain areas where particular types of rock are more prevalent. It is always advised to use a native stone in your work, primarily because it fits the look of the landscape. It is odd to see a startling piece of black granite in the middle of a cactus garden in Arizona, for example, but a carved piece of sandstone would perfectly accent the surrounding garden materials.

Certain regions of the United States are rich in natural rock, particularly those with mountain ranges. In most of Kentucky, dry stone rock walls and stone-built homes are reminders of the state's pioneer past and are considered national treasures. You can find stone by walking your property and digging up usable stone, or you can visit local quarries or stone yards. You can find out if your city or town has a stone yard by checking the phone book or by searching the Internet for the closest stone yard in your area. It is more beneficial to purchase stone from a stone yard simply because you can choose your stones for their already formed shapes, colors, and sizes, and you can have them delivered, which is advantageous if you do not own a truck or have access to one. But, if you are building a rock or water garden and have usable stone on your property, it is perfectly acceptable to use native rock because it blends with the surrounding landscape.

A caution exists in using fieldstone: You cannot remove native rock from state parks or preserves without a National Forest Service permit or Department of the Interior permit, and you certainly cannot use rock found on other people's property without their consent. Some builders and contractors welcome the removal of stone from specific building sites, but you still need to get permission from the landowner before removal.

In order to obtain a permit for gathering rock, which is considered a forest product, you need to contact your local state office for the National Forest Service. The website for information is **www.blm.gov**, or you can call the toll-free number for the U.S. National Forest Service at 1-800-832-1355. You can access individual state offices by going to their website at **www.blm.gov/wo/st/en.html**, calling 202-208-3801, or writing the main office at:

Bureau of Land Management
Washington Office
1849 C St. NW, Rm. 5665
Washington D.C., 20240

More About Fieldstone

Fieldstone is defined as just that: Stone that occurs naturally in fields and is used as building material. Gathering fieldstone can be a laborious process that entails many trips back and forth from creek banks and mountainous areas. The National Forest Service imposed a fee for stone collecting, which allows you to gather a specific quantity of stone over a certain time from national forests and for personal use only. They impose a fine for not getting permission to collect stone and rock from public lands. Gathering from streams, using heavy machinery, and gathering for commercial use and resale is prohibited. What is not prohibited is collecting rock and stone from mountains and hillsides, along riverbeds but not in the stream itself, and pasturelands. Backhoes and bulldozers, commonly used to quarry rock, destroy public lands, which is why they are not permitted.

Fieldstone is used like any other stone because **dry stacking**, which refers to stacking dry rock on top of each other in such a way the stones interlock and stand firm without the use of mortar, can hold fieldstone in place. Fieldstone is commonly used in building outdoor walls and fences; structures, such as springhouses, grottoes, or residential homes; and oftentimes fireplaces. Using fieldstone is trickier than using quarried stone because the surfaces are not cut, and they require shaping. But, the weathered look of the stone provides one of the reasons people prefer using rocks they find in nature as opposed to ones they purchase from a quarry. Other benefits to using fieldstone include that it has already aged to the color it will be when used; it has softer, rounded edges; and best of all, it is usually free, or cheaper, than purchasing quarried stone.

This stone wall was made using dry stacking techniques.

You should choose fieldstone close to the shape you will need so you have minimal work to do to the stone once you move it. Limestone and sandstone both occur in layers, and finding a few loose stones usually means there will be plenty more nearby. You can easily load the bed of a truck with stones, and after a few trips, you may have enough for your project with minimal or no financial cost.

When gathering fieldstone, always remember to exercise proper caution. Never climb up or down wet slopes, never try to lift overly heavy stones, and take your time when gathering stones to avoid straining your muscles and to not put yourself or others in danger by rushing about. Your backyard barbecue pit project is certainly not worth broken bones. It is also good safety practice to travel with someone else in case a problem occurs and to help you gather and lift stones.

Roadsides near farming communities also contain fieldstone. Farmers often plow their fields and remove the resulting rock and stone by throwing them over a fence and away from their crops. You can gather fieldstone from alongside creek beds but not from within the streams, as it is illegal in all states. Although the rocks in streams are tempting to remove because they are large and rounded by the constant washing of water, you cannot use them in your projects. Whenever a landowner prepares property for building a home, you can usually ask the landowner for permission to load up some, or all, of the upturned rock.

Example of a house that was made out of fieldstone.

Good sources of **recycled stone** include abandoned or burnt brick or stone homes and chimneys. You can use these stones or brick in conjunction with other fieldstone. Many old chimneys and structures built before 1900 and the advent of Portland cement used a form of mortar called **lime-and-sand** mortar, made from dry lime and sand and then wet with water to make a loose, adhesive grout. This grout, over the course of a few years, became quite brittle, and the lime helped corrode the soft brick. Most of this lime-and-sand mortar is easily dug out away from the stone so you can easily reuse the stone itself. Again, always get permission from landowners before gathering. Do not assume the landowner will not care because the stones have been on vacant property for years. It could be a National Historic Register structure, which carries a huge fine to the hapless stone gatherer if meddled with.

Tips for gathering fieldstone

Although you may think gathering fieldstone simply requires picking up a few rocks and tossing them into the back of your truck, certain things can maximize how efficiently you gather stone, preventing unnecessary trips and saving time.

1. Choose stone by shape

When gathering stone, look for flat surfaces and square or rectangular shapes. These stones will form the corners, overlapping the other stones on two sides and providing a flat base for additional stones. These stones will also serve as **arch stones**, which are stones that form an arch, if you are including one in your building construction. These kinds of stone lay easier and require little cutting. Ideally, fieldstones are used without mortar, and they should fit as evenly and tightly together as possible. This does not mean you should always disregard a stone because it is not perfectly flat or square. Many dry stone fences and walls are made with beautiful, flat stones set in a vertical or horizontal fashion as **capstones**, defined as stones that finish off the top of a wall.

2. Gather more than you need

This is a standard rule of thumb for stone gatherers: More is better. You never know if you will need more or less, and it is always better to have excess stone so you will not have to stop in the middle of your project and go in search of more or wait for a delivery from the local stone yard. Whereas carpenters have a measure twice, cut once rule, stone gatherers have a 50 percent overage rule, meaning you should always gather 50 percent more rock that you think you may need to complete your project. You can always use this rock later in another project. If you have a stone yard in your town, you may possibly sell your overage to them, and they will load it and carry it off for you. As long as you gathered the rock for personal use initially, if you have an overage, then you can dispose of it in any fashion you see fit.

3. Be kind to the Earth

When you dig out stones from public or private lands, fill in or cover the areas where you dug with leaves and brush to keep the area from washing out in heavy rains. Be respectful of the property of others if you gather on their land, and always try to fill in the spaces where you dig with earth and brush, leaving the scenery intact and just as beautiful as you found it. Remember: You must always have permission to dig on any public or private land.

Quarried Stone

Quarried stone is stone mechanically extracted from the earth rather than exposed due to natural erosion or the preparation of building sites. It is stronger than fieldstone because the surface of the rock has not been exposed to weather for a long time. As stone naturally ages, minute cracks allow water to seep into the rock's interior, which freezes, and eventually these small cracks enlarge and lengthen. The stone will then break into smaller stones and, over thousands of years, it eventually becomes sand or grit. With quarried stone, this natural process has not happened, and the rock is harder and denser than fieldstone. There is also a caveat to quarried stone: In most cases, the quarried stone has at least one flat side or more. This is essential when building structures to ensure a level and even surface.

Most quarried stone is sold through stone yards, which can be a blessing or a curse. If you are looking to build a dry stone structure, you are rarely allowed to pick and choose the stones you want because they are sold on a pallet with plastic wrap to keep the rocks and stones intact, and these pallets do not always include the best stones for the type of project you are working on. Most stone yards cater to masons who use mortar for building their projects. You can certainly mix commercially quarried stone with fieldstone to make your purchases more cost effective.

Faux stone

The Southwestern Native Americans are credited with the first fabrication of **faux stone**, which was native soil and clay formed into small bricks. They used discarded vegetable matter, such as corn shocks and wheat chafe, to strengthen these bricks, which is commonly referred to as **adobe**. Because adobe bricks are man made, they are considered the first faux stone. In modern times, the availability and diversity of faux stone has come a long way. Commercially made faux stone began with a product called Z-BRICK®, one of the earliest manufactured stones. Now, many faux stones exists, mimicking every natural stone and texture with every color imaginable. Even entire faux stone panels exist, making it even easier to install a stone-looking wall in your home.

The Zygrove Corporation, the manufacturers of Z-BRICK, was the first company to manufacture and market this type of product to the public for use inside and outside the home. Originally, making your den or kitchen look like it had brick walls provided the most popular use for Z-BRICK. Z-BRICK now manufactures more natural-looking veneers and has a veneer stone specifically for outdoor walls in exterior foyers and patios. It has been one of the best selling products for this purpose since its introduction in 1955 and continues to be in demand today. One of the components of Z-BRICK is **vermiculite** — a natural mineral formed with a high concentration of clay, which expands with heat. Vermiculite is used in commercial and residential insulation and is used to add air pockets to garden soil. It also makes Z-BRICK stone and brick lightweight and natural in texture. According to the website for Z-BRICK (**www.z-brick.com**), the main difference between a Z-BRICK and an all clay-based brick is Z-BRICKs are lightweight, dense, and thin instead of brittle like natural clay brick, and they are more cost efficient than masonry brick.

Currently, hundreds of companies manufacture faux stone and faux stone panels with lifelike replicated textures. You can order panels that slide open, faux rocks for waterfalls and other outdoor uses, and even **replicated stone**, which is the new terminology for faux stone, in bendable sheets for encircling round objects. Manufactured stone panels can vary greatly in price from one dealer to another, and you must remember the golden rule of retail: You get what you pay for. Replicated, or faux, stone can be just as visually appealing as natural stone, but many of the cheaper products do not have the intricate colorations that natural rock offers. Therefore, if you are considering using faux stone in any veneer work, then weigh your options carefully and choose a stone that will give you the accent you are looking for at a reasonable cost, but do not always go for the cheapest option if you can afford it. In the long run, it is always better to spend the money on a higher quality of faux stone because it will reward you with many years of durability and good looks. This book will focus on using natural stone, rather than faux stone in various projects.

CHAPTER 2:
Working with Stone Tools

Companies manufacture and sell more high-tech tools to consumers who want to work with stone than ever before. Earlier stonemasons had only chisels and hammers to use when cutting and shaping stone. But, even with all the new tools available, many masons still pick up a hammer and chisel and form stone by hand, just as it was done centuries ago because it is less complicated and more economical than obtaining expensive and complicated machines. This chapter explores the basic tools needed to shape and form stone and rock, as well as how to lift and transport stone. Each step will explain how to properly divide, split, or cut raw stone for dry stone or mortar masonry and will describe power tools available to modern stonemasons.

Lifting, Loading, and Transporting

Before you can begin building with your stone, you need to move it to the project site. Depending on the sizes and weights of your stones, you may only require a pickup truck with a large piece of plywood in the bed to protect against scrapes and dents. If you are trying to transport large rocks that are too heavy or bulky for you to lift safely on your own, you will need something a bit more heavy duty, such as a heavy flatbed truck or bulldozer. It is best to leave the stones weighing a ton or several tons to the professionals who move stone with heavy construction machinery.

You will also want to take your hammer and chisel into the field with you if you are gathering your own stones. This allows you to shape rocks that are easy to cut. You can determine the nature of rock by chipping or cutting it on the site to see the interior of the stone, as well as to determine how hard or dense the stone is. For example, using a hammer and chisel to break a stone in half will allow you to see the interior of the stone where you can determine if it is sandstone, fieldstone, or granite. You can also shape the stone on site to prevent taking more weight than you need to the project site. And, of course, you can leave the small rock you chip off the larger ones when you shape your stones to naturally degrade and become part of the soil.

Lifting anything of weight or bulk is better if you have two or more people because you want to avoid back or joint injuries. If you decide you really want to take off into the woods alone to look for stone, and you find some heavier rocks you just cannot resist taking back, you will need to apply some basic safety rules and remember a very important fact: Leverage is your friend. Using leverage will greatly increase the weight you can move or lift without injuring yourself. Using the crowbar to dislodge stones, rolling the stones along a makeshift roller belt by using polyvinyl chloride (PVC) piping, and making a hoist are also ways you can move stone without too much effort.

If you need to lift a heavy stone by yourself, always implement the following steps:

1. Rest the stone against your upper thighs to prevent lifting the full weight of the stone with your arms.
2. Bend your knees, keeping your back as straight as possible.
3. Slowly stand and lean back slightly so that your legs do most of the lifting and not your back.
4. When you walk to your vehicle (or the job area) with the stone, watch your footing. If you stumble, release the stone, pushing away as you release so it does not fall on your toes.

It is always better to have another person help you lift and transport rock and stone, but if you cannot find another person to help you, you can use certain methods to help you achieve your collecting goals. Realize your own limitations, and if you find some rocks you want to collect but they are too heavy or too bulky, make a note of their location and come back when you can find someone to help. In the meantime, here are six methods that can help you lift the heavier stones and transport them.

Board and piping

Lay a large, flat board, such as a sheet of plywood, on top of several metal or PVC pipes. Set your stone on top of the makeshift sled and roll, removing and replacing the pipe from the back of the sled to the front of the sled as you go along. If you have a large stone slab to move, you can eliminate the plywood if you have large enough PVC pipe — it is best to use a pipe with a 4-inch diameter. You do not need another person to help you transport stone using this method, but it does make it easier.

A **sledge** is a sleigh or sled with a flat base with handles and runners on the base to help with mobility. It usually has a pull rope or chain in the front for steering. Using chain instead of rope is recommended so you can pull the heavy stones without worrying about the rope breaking.

Example of a sledge.

Hand truck

A sturdy **hand truck**, also called a dolly, is a highly valued tool for anyone who has to move heavy objects. A dolly is made up of a small cart that has a frame with two low wheels and a ledge at the base. It has handles at the top and is often used to move crates or other heavy objects. You can position heavy stones on the hand truck by gently raising the

handles up to transport the stones without having to secure them. Always put larger, flatter stones on the bottom, and if you are moving very heavy stones, then never move more than three at a time to prevent damage to the hand truck.

Tire sled

You can make a functional sled from an old tire and a tarp. Simply tie a chain around one end of the tire and insert the tarp into the rim hole in the tire. Place your stones into the center and drag the tire by the chain.

Use a makeshift lever

Using a heavy board, a 2-by-12-inch oak or other hardwood plank is recommended, as a lever can help you lift stones out of the ground and onto a sled or hand truck. You can also use a plank to prop against your vehicle and slide or tumble heavier stones up the plank into the bed.

Wheelbarrow

This is another invaluable tool for transporting stone. You can lay it on its side and slide the stone inside. Turn the wheelbarrow back upright and push it to the location. Use the same methods for unloading the stone.

Trucks

In the event that you become a die-hard rock gatherer or you choose to move on from smaller stone projects to building homes or other structures from stone, you will probably want to invest in a standard size pickup truck with manual transmission — allowing you to downshift and save on fuel and brake pads — and an **electric winch**. This device requires a heavy-duty cable with a hook powered by the vehicle's battery and is used to pull rock out of difficult and/or wet areas. A four-wheel drive truck will cost more in fuel, but it may allow you to navigate safely down to creek beds and through steeper, mountainous ranges. You may even look to have a hydraulic lift installed in the back of the truck, similar to a tow truck. If you progress to installing this type of heavy-duty machinery, then you will need to take special consideration of the truck itself and the loading capacity of your mounting bolts and braces, axles, and wheels. It is also important to remember that once you load your stones, you will have to unload and possibly transport or carry the rocks up a scaffold if you are building a fireplace or chimney. This is why you may want to enlist the help of others to help you lift the stones. Another stonemason's rule applies here: It takes just as long to lay a small stone as a large one. Although it looks good to alternate large stones with small ones, small ones may be all you can handle.

The Essential Tool List

Each project outlined in this book has a specific materials and tools list. You may need certain tools for one project that you may not need for others. For example, if you are building a veneer stone wall over concrete block, you will need masonry ties, whereas if you are building a dry stone wall, you will not. Therefore, pay close attention to each project's materials and tools list to avoid buying more than you will need for that particular project.

All stonemasons — regardless of experience level — need various tools to get the job done right. When you start to purchase chisels and hammers, you will find many different types of tools made from steel, carbide, and titanium, and it appears to be a personal choice as to which is better among stonemasons. But if you use power tool bits when working with hard rock, such as granite, carbide will outlast steel. The tools you may need are listed here, as well as a short description of their primary functions.

Gloves

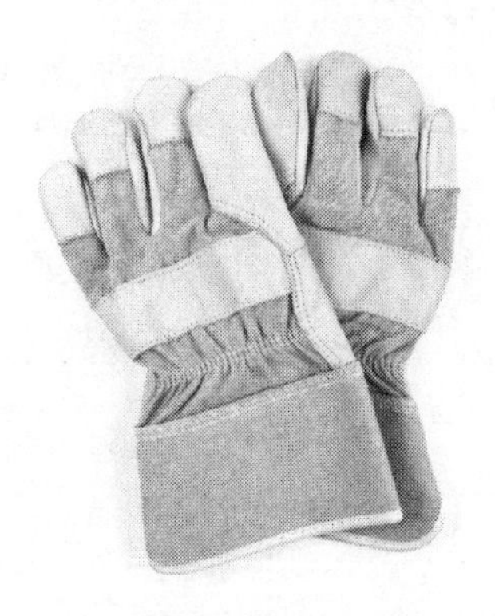

A heavy-duty set of gloves, preferably leather, will save your hands from scrapes and cuts. It is best to have several sets in different sizes scattered around in case someone sees you laboring away and decides to help.

Also, when you mix mortar and cement, you will need specialized gloves because mortar and cement have caustic qualities that will burn the skin off your hands. Cement contains a chemical called hexavalent chromium, which causes allergic dermatitis. Prolonged exposure can cause severe dermatitis, which is an inflammation of the skin that causes severe itching, flaking, and burning. Gloves made from rubber are best for using when mixing cement.

SIDE NOTE: If you work with Portland cement on an extended basis (including grout), you can protect yourself from caustic burns or dermatitis by following these safety tips:

1. Always wear gloves when mixing, troweling, or grouting with any cement-based product.
2. Wear long sleeves and long pants or wear rubber boots.
3. Wash your hands frequently with running water from a hose or sink and do not use abrasive cleansers containing limonene — these cleansers usually have a citrus odor and are used for removing oil and gasoline.
4. Remove your gloves from the fingertips and not from the cuff.

Safety Goggles

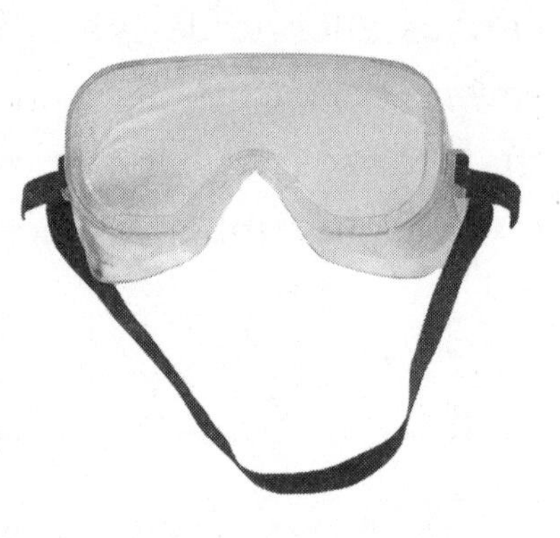

Protective eyewear is essential on job sites that have anything to do with flying rock, blowing sand, or cement dust. The importance of wearing goggles cannot be stressed enough. One small splinter from a chiseled rock flying into your eye can cause permanent blindness. Even if you wear glasses, slipping a pair of safety goggles over your glasses protects the lenses. Several types of safety glasses exist, as well as tinted safety glasses to help protect your eyes from the sun when working outside for an extended time.

Buckets

Buckets are useful for many reasons, but you will need them more than you ever realized as soon as you begin building with your stones. You can mix small amounts of mortar in large, empty commercial paint buckets, load smaller stones in the buckets to easily transport them, fill them with water for washing up tools, and even store hand tools in them when the day is finished. You can purchase these buckets at home hardware stores for less than $3 each.

Hoes

Hoes are long-handled tools with a flat blade attached at a right angle used for weeding, cultivating, and gardening. When building with stone, it is a perfect tool for mixing mortar, dislodging small rocks, leveling ground, and weeding if you are building an outdoor structure. There is a hoe specifically designed for mixing mortar and cement that has two large, round holes in the blade that allow mortar or cement to gush through while mixing, but it is not entirely necessary to have this specific hoe. A simple garden hoe will achieve the same results. It is important to wash the hoe off with running water after each use or the cement and mortar will set up on the blade, and you cannot use it again.

Shovels

Shovels allow you to dig footings, move earth, and add mortar mix to foundation footings. The best shovel for these applications is a heavy-duty shovel with a flat blade. Newer shovels have reinforced handles made from construction-grade plastic that will not rot or weather.

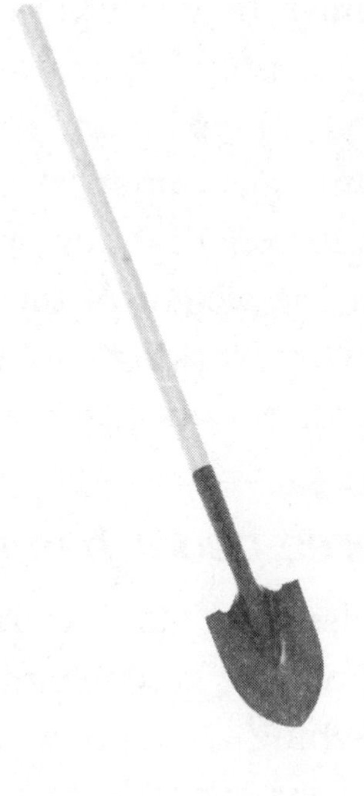

Crowbars

Crowbars are tools used for leverage and demolition, and you will be doing both when working and building with stone. A crowbar is a straight bar of iron or steel with the working end shaped like a chisel and often slightly bent and forked. Originally called crows in Colonial America because the split end resembles a crow's foot, they come in many different

sizes and weights. They are also referred to as prybars, jimmys, or jimmy bars. Using the crowbar as a leverage tool, you can pry larger stones from the earth. They can also be used to help lift and roll large stones onto a sled or wooden plank for transporting. There are lightweight crowbars made from titanium, which is a nearly indestructible metal, but the weight of the crowbar works in your favor when you are working with rock so choose one made from steel because it offers more weight than a titanium version.

Three-pound hammer (sledgehammer)

These are also referred to as 3-pound sledgehammers or mashing hammers. Seasoned masons sometimes make their own from heavy metals like truck axles and use hardwoods wedged into holes drilled through the centers of the metal. The commercial 3-pound hammers with tempered steel and fiberglass is an acceptable substitute to making your own. This tool is used for striking a chisel when splitting stone, but you can also use a steel mallet for striking chisels.

Mason's hammer

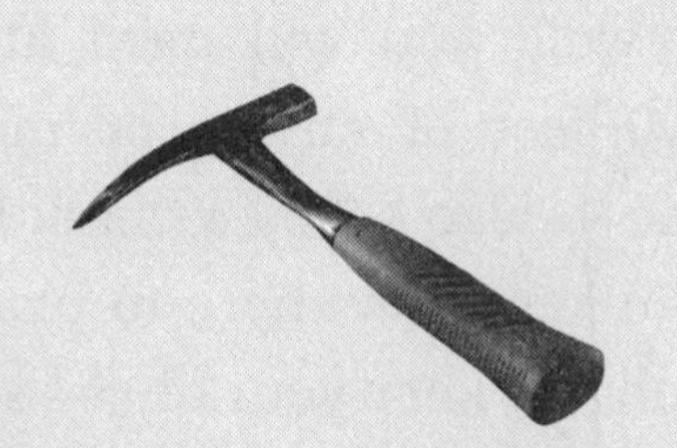

This tool has one end shaped like a common hammer and the other end fashioned with a curved chisel. The mason's hammer is used for splitting block or brick with the chiseled end by creating light blows all the way around where you want the stone to split. You then use the hammer end to sharply tap the stone, causing the stone to split cleanly. You can clean any remaining edges or burrs using the chisel. This is one of the tools you will use the most as a stonemason.

Levels

Any project you build will need to be level, or as level and plumb as you can get it using rock. The longer the level — a carpenter's level, which is 4 feet long, is preferable — the easier you can determine whether your project is on the right track. For example, if you are building a connecting wall, you must have a level surface on both sides in order to have the capstones level. Make sure the bubbles that indicate alignment are both vertical and horizontal.

String

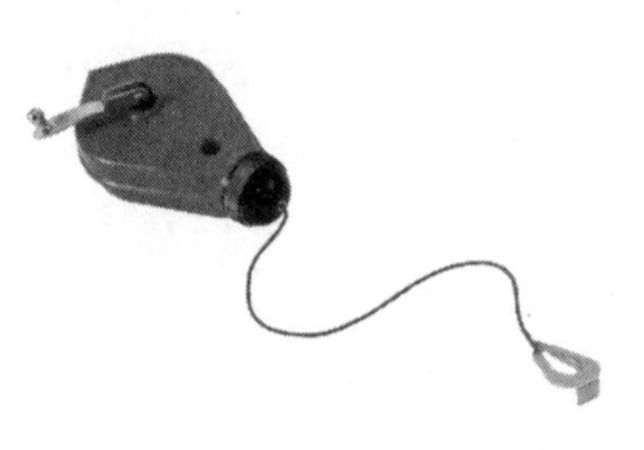

If you are building a structure or wall, you will need at least 100 feet of string, or mason's line — also called a chalk line — to determine where to place the wall and keep your lines straight. Natural string in cotton or jute keeps knots taut better than nylon, but nylon weathers better if you are working with an outdoor project that may take weeks or months to complete.

Charcoal pencil

This type of pencil is readily available in most hardware stores and art supply stores. It makes a clean black mark on rock and stone for measuring chisel lines and for drawing out areas to carve, yet washes off easily with water.

Stakes

Building walls and pathways can employ the use of small wooden stakes for marking out the areas the walls and pathways will go. Stakes help keep your pathway or walls from veering off course.

Measuring tape

You need a metal, retractable tape to measure the lengths and widths of building stones and walls and to determine the length of an area you will build upon. A measuring tape

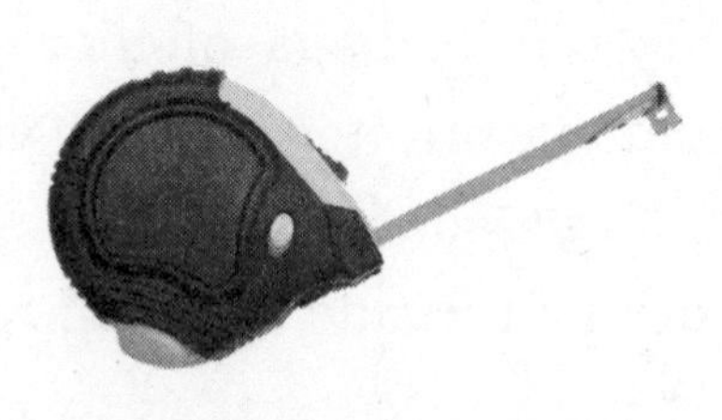

with a 24- or 30-foot expansion should suffice for the projects described in this book.

Wire brushes

Wire brushes are used to clean moss and debris off fieldstones, as well as to clean out cracks and crevices in walls. They will also help clean old mortar off brick and stone. You can also

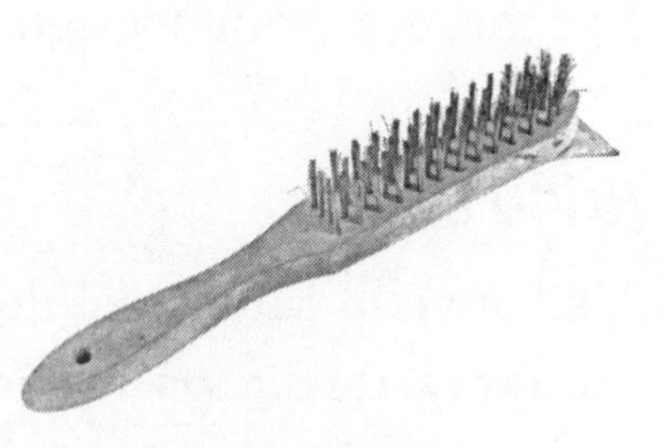

use wire brushes to texture grout to resemble original grout when doing restoration work.

Scoops for mortar and cement

You can use shovels to get dry mix out of bags of mortar and cement, but it is easier if you use a large scoop because shovels rarely fit inside the bags.

Mortar pan (mixing trough)

These are large metal or plastic bins that you mix cement in. You can also use your wheelbarrow, but if your wheelbarrow is stacked with stone, it is always nice to have a separate mixing trough.

Hand spade

A spade is a hand tool designed primarily to dig or remove soil and to break clumps of earth. It is narrow and pointed, much like an arrow, with a short handle.

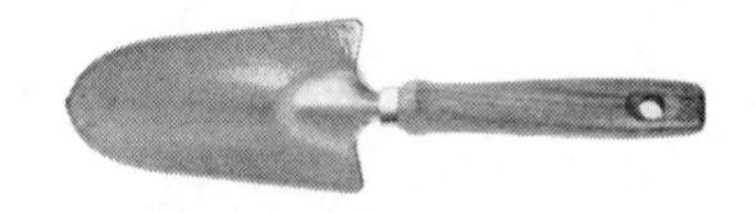

Mortarboard

A mortarboard is a large piece of hand-held wood that excess mortar can be thrown on as you work, much like an artist's oil palette. You can reuse excess mortar in your building projects as long as it has not begun to cure and dry.

Stone grinder attachment

There are hundreds of grinding bits and attachments made for use in the stone-cutting industry. The most popular bit for stonework is a diamond-gritted grinding cup, which is an attachment to a drill that you can use to core out basins and smooth rough stone edges. These grinding bits are attached to an electric or pneumatic drill like any other bit. Other popular grinding bits are:

- Grinding wheel — used to remove excess stone when polishing.
- Machine chisel with finishing blade — used for general carving and sculpting.
- Machine chisel with roughing blade — used when removing heavy stones is necessary.
- Four-point tooth chisel — used to aggressively remove excess rock.
- Splitter or cape chisel — used for fine-tuning lines in rock and stone. It is also used to remove old mortar in preparation for pointing. Pointing is the process by which old grout and mortar is removed from brick structures and replaced with new grout and mortar without tearing down the structure itself.

Stone chisels (also called blocking chisels)

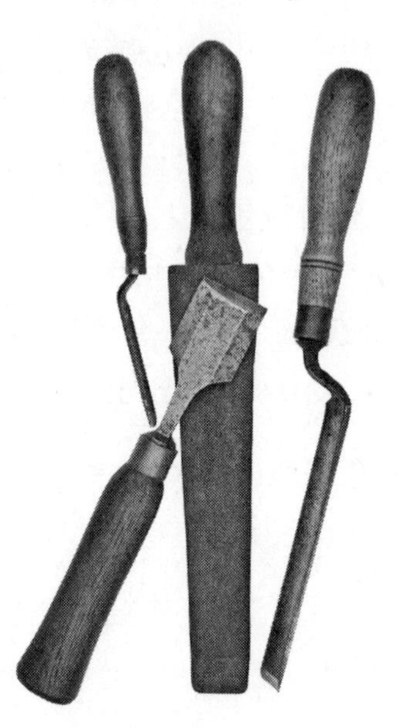

Stone chisels are large metal wedges used to create splits in stone. They come in many different shapes and sizes but are all designed to split stone. It is recommended that you have one wide chisel — 8 inches wide, also called a brick chisel; a medium-sized chisel — 4 inches wide; and a small chisel — 1 inch wide — for smoothing jagged ends or to fashion tight cornerstones. Small chisels are also used for removing old grout and mortar.

Pick

The pick closely resembles a mason's hammer but has a spaded, pyramid-shaped point on one end and a curved pointed end on the other. It is used to find rock under the soil and to dig with.

Rake

A garden rake, preferably metal, is useful for clearing excess stone debris and helping create a level surface on which to lay stone. You can also use this tool to clean around the finished structure.

Earplugs

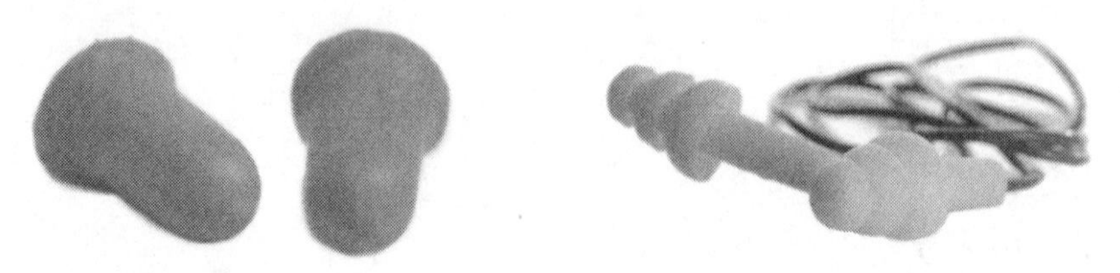

If you will use power tools, earplugs are essential to protect your hearing.

Masonry ties

Masonry ties, also called brick ties, are small metal reinforcing clips used in building walls to help tie veneer stone to concrete walls. They are used whenever and wherever structural strength is needed on brick or veneered walls. They are normally placed into the wet mortar every second or third block.

S-hooks, anchor bolts

When building forms and other tools, such as a ratchet hoist for moving larger rocks over stream beds or installing stone bench seats, you may need large S-hooks and anchor bolts. S-hooks will hold the chains for the ratchet hoist, and anchor bolts will keep the legs of the hoist together, as well as support braces for stone pillars.

Garden hose

Water is necessary for mixing cement and mortar and for cleaning tools after you use them. Having access to a garden hose and spigot rather than carrying water in buckets is best. You will also need access to a garden hose if you plan on creating a water garden.

Trowels

Trowels are flat-bladed, pointed hand tools used for leveling, spreading, or shaping substances, such as cement or mortar. Trowels are also used to apply mixed mortar, and the two trowels you should have in your tool kit are a pointing trowel and a trowel with a blade roughly 8 to 12 inches wide. The 8- to 12-inch trowel is good for applying mortar to flat, level surfaces, and the pointing trowel is a finishing trowel good for cleaning mortar joints and to finish work on the exterior. Choose trowels with short handles because manipulating a long-handled trowel can be hard on your joints after a long time. Also, choose a trowel that has a thick metal band where the trowel blade joins the handle because most of the stress occurs here.

Sponges

When working with grout, you will need to use grouting sponges, which are large cellulose sponges used to wipe excess grout off the rock. They are 6 to 8 inches long and at least 4 inches thick, but you can cut them to whatever size you feel comfortable working with. It is important to keep sponges clean from debris while working with them and important to rinse them often to keep mortar and cement from collecting in the sponge and hardening.

Power tools

To cut harder stones, such as granite and basalt, you need tools that use an external power source in the form of an air compressor, which uses air pressure to run drills, cutters, grinders, and splitters. Many of the power tools used in commercial applications are huge and not practical for the beginning stonemason. But, companies have manufactured simple and affordable power tools for the beginning stonemason to make cutting through harder stones easier.

AIR COMPRESSOR

All of these power tools operate off an air compressor and a 1-inch or 3/4-inch "D" or "B" type pneumatic tool base. The "D" and "B" designation is based on size variations in drill bits. The air compressor, which is a large motor encased in metal, also has an air hose and a quick connect for various bits and chisels. You can purchase air compressors at any hardware store. The size of your air compressor should depend on how hard the rock is you will carve and shape. For instance, if you are dealing with limestone or sandstone, you will not need much power. Working with harder rock, such as granite, requires a larger air-compression unit. Also, if you intend to do quite a bit of shaping, sanding, and carving, it is best to invest in an air compressor with a higher horsepower, which will provide a better long-term investment and fewer repairs. A five-horsepower air compressor should supply enough compression for working with either type of pneumatic tool base. You can use several different interchangeable chisels for this type of work and most are finishing chisels for cleaning up rough cuts and carving. You can also use diamond blades and grinding stones for cutting edges into stone and for keeping your chisels sharp and free of metal burrs.

TYPE "B" PNEUMATIC TOOL BASE

The type "B" pneumatic power tool has a shorter piston stroke, offers less vibration, and is quieter to operate than the type "D" power tool. The type "B" is better suited for light use and softer stones, such as limestone and sandstone.

TYPE "D" PNEUMATIC TOOL BASE

The type "D" pneumatic power tool is specifically designed for harder rock and provides more vibration and noise than the type "B." If you are going to be cutting primarily granite and basalt, invest in the type "D" tool base.

A note on sharpening your chisel bits when they become dull: When you sharpen chisels using a grinding stone or blade, you can cool them in a bucket of water as you go but only if you are sharpening steel. If you sharpen carbide-tipped chisels in water, the chisel head may crack because of the rapid temperature change. There are two schools of thought in choosing steel over carbide: Carbide tips are considered harder and more durable than steel and do not tend to develop the "mushrooming" effect of steel over time. For example, as you continue to hammer on a chisel, the top will slowly flatten and form a lip over the chisel top. This effect is called "mushrooming." This can be very dangerous as the sharp edges on the mushroomed lip can break unexpectedly and fly off. It is important to keep these edges ground off with a grinding bit. A grinding bit is a drill bit made from diamond or hard silica that will grind metal. You have to let carbide tips cool down from time to time as they do get hotter with continued use, and you cannot cool them in water as you can with steel.

BUSHING TOOLS

Bushing tools are specific tool bits used to **finish-carve**, which is the term used to describe the final phases of carving on a stone piece and adding texture to stone, such as cross-hatching or facial details in sculpture. These tools are the four- and nine-point chisels, the carver's drill, and the cup chisel. None of these bushing tools are used in any of the projects listed in this book and are normally used by stone sculptors in the completion of their art. It was important to mention them in the event you choose to carve lettering into stone.

Additional tools

Other tools or accessories you may find you need are:

STEEL-TOED BOOTS

For safety reasons, these feet protectors are absolutely valuable to any rock gatherer and stone builder. Accidents happen; therefore, if you have a pair of steel-toed boots, you will be exceptionally thankful should a rock of any size come into contact with your foot.

MOVING STRAPS

Moving straps are densely woven nylon or rubberized straps designed to fit under or over heavy equipment or furniture. They are useful in moving stone because they conform to any shape object and are strong.

TILE MASTIC

Tile mastic, a premixed tile adhesive used in the construction industry, looks like a joint compound and sets up slowly, allowing you to reposition easily. Tile mastic is an all-purpose adhesive but is used primarily on flat surfaces.

STUD FINDER

A stud finder, also called a stud detector or stud sensor, is a hand-held device used to determine the location of wood and metal framing studs used in light-frame construction after you have installed the walling surface.

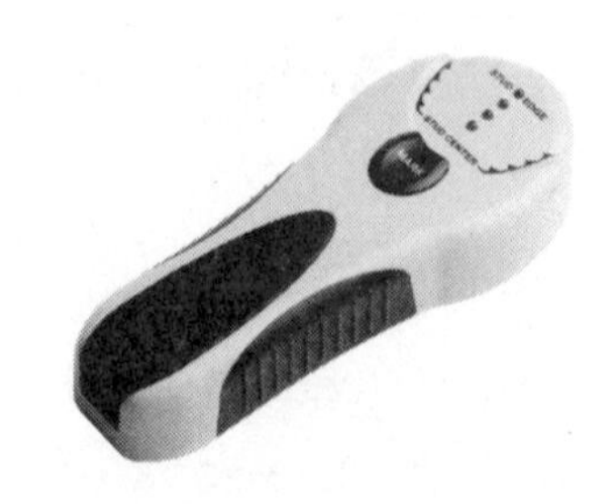

METAL LATH

Metal lath is a sheet of metal that is slit and drawn out to form openings. It is used as a plaster base for walls and ceilings and used as reinforcement over other forms of plaster base.

LEATHER APRON

A leather apron is more durable than a standard denim or cloth apron and will protect your clothing while carrying or working with stone, cement, and mortar.

GRADE 1 LADDERS

The American National Standards Institute determines the "grades" of ladders. These grades are listed on the next page.

You will need a minimum Grade 1 ladder if you do any climbing with your stone project. You may also need scaffolding if your project requires a height of more than 6 to 8 feet.

Classification of ladders		
Grade	**Capacity (lbs.)**	**Comment**
IAA	375	Special duty
IA	300	Extra heavy duty
I	250	Heavy duty
II	225	Medium duty
III	200	Household use

Grade 1 Ladder.

HARD HAT

You may find that wearing a construction hard hat when working on projects above your height level or when on ladders or scaffolding is an excellent extra safety procedure to protect your head from falling or dropping rock and stone.

CIRCULAR SAW

You may need a circular saw to cut wood when making a form or brace in stonework. You can also use a hand saw or have your local hardware store cut the lumber to length as requested for a minimal charge.

Battery powered circular saw.

JIGSAW

This hand-held power tool is primarily used for cutting patterns, usually circular, in wood because the blade moves up and down instead of across, allowing for detailed cuts.

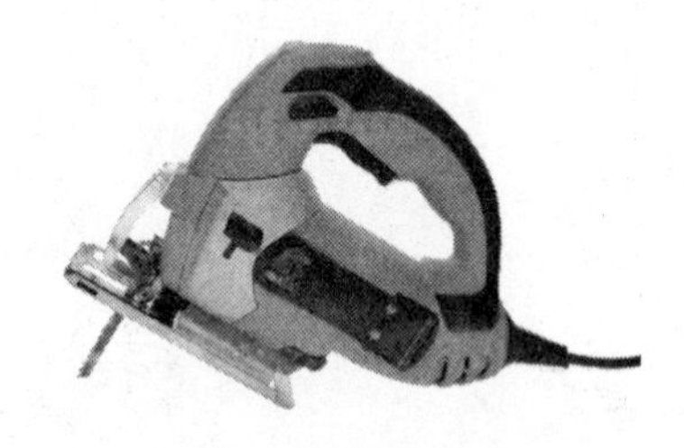

METAL GRID

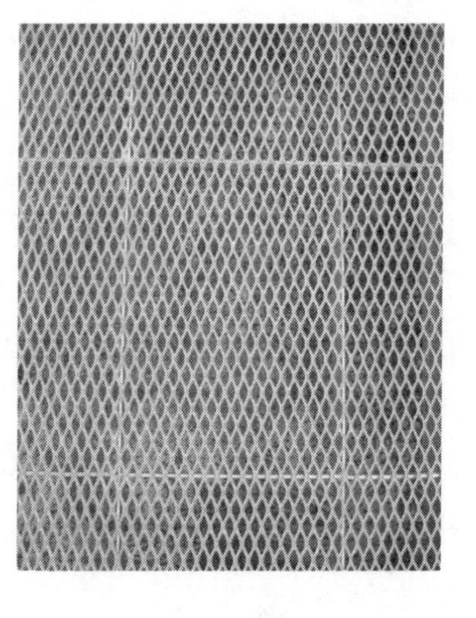

Metal grid is simply a metal cooking grill, such as the one made for portable or free-standing cooking grills. If you are making a barbecue pit, you will need a manufactured metal grid, or grill, on which to place your food when you cook. You can buy these grids wherever replacement parts for portable grills are sold.

RUBBER LINER

Here is an example of rubber liner being used in a backyard pond. Photo courtesy of Douglas R. Brown.

If you are building a water garden, you will need a rubber liner for the pond to hold the water if you do not use a fiberglass or plastic premolded form. You can purchase these liners in rolls to accommodate any pond size.

POINTING TOOL

If you are doing repair work to brick, you will need to add a pointing tool to your tools list. A pointing tool is also called a pointing trowel, which is a small, short-handled trowel used to replace mortar between brick.

SHOCK-ABSORBING GLOVES

Gloves are always necessary, but when using power tools for an extended time, you will need to invest in shock-absorbing gloves. Shock-absorbing gloves, also called impact-resistant gloves, usually have a mesh back for air flow and a gel-filled palm designed to reduce stress and fatigue when operating jack hammers and pneumatic drills used in shaping stone. Some gloves have additional leather padding across the palm as well. Leather gloves alone will not provide the protection needed for impact tools.

FILTER FABRIC

If you are building water gardens, you will need to put an underlayment of filter fabric down to protect against weeds, rocks, and debris and to protect the rubber liner from tearing or abrasion. If you are building a retaining wall, it is advised to put a strip of filter fabric underneath the top capstone and along the embankment edge to help prevent debris and soil from filtering into the stones and causing a possible weakened area where the wall may crumble.

DUST MASKS

Dust masks are essential when using any kind of power towel for grinding, drilling, or sculpting rock or stone. Many natural stones contain trace elements of minerals not good for you to breathe into your lungs, such as lead, zinc, and calcium carbonate.

While you can purchase all or some of these tools at your local hardware store, you can also purchase them on the Internet. One source is Master Wholesale®, Inc., based in Seattle, Washington. The company's website is **www.masterwholesale.com**, and the phone number to request a catalog is 1-800-938-7925. Another source is Trow & Holden, based in Vermont, and the company website is **www.trowandholden.com**.

Now that you have the tools you will need for stonemasonry, it is time to start learning to use them. The beginning steps to your projects usually involve learning how to cut and shape the stones you have purchased or gathered.

Cutting and Shaping Stone

The idea of cutting and shaping stones might seem like a task better suited for Paul Bunyan, but once you have read the following sections explaining the use of tools and techniques you will need to shape the stones for your project, you will be surprised at how easy it actually is. Most tools used for cutting and shaping stone use leverage and weight and require little effort on your part.

To know how you want to shape or cut your stone, you must first determine what stone you will use and how you will use it. For example, a barbecue may be square, which would require you to have square-cut cornerstones if you are using fieldstone or sandstone. If you wish to build a circular fire pit, then you do not need cornerstones or shaped edges, but you will need **shims** and **wedges**, which are small shards of rock used to fill in and help level rock structures when using unshaped rock.

In order to successfully begin and execute any endeavor, planning is a vital step. It is impossible to know how much stone you will need for your specific project, as the requirements are determined by how wide, how long, or how deep something will be. You can approximate the amount of stone you will need and the amount of mortar initially, but that may change drastically as you build. Start with the basics, and add as you go.

Using a hammer and chisel

No matter what project you begin, if it requires structure, such as a wall, bench, or barbecue, you need somewhat square or rectangular stones to work with to allow the stones to fit properly to provide stability and strength.

If you are out in the field cutting and shaping stone, follow these simple steps to cut rounded corners or jagged edges off rock:

1. Lay the rock on padding, such as old blankets or pillows, to prevent the stone from sinking into the ground. If possible, bring a sturdy wooden or metal table with you, which should be roughly the height at which your knuckles come. This will give you enough leverage when you strike the chisel to force a cut in the rock and will keep you from bending over to work at ground level. If you cannot bring a table, at least bring a wooden box or stool to sit on while breaking the rock at ground level to prevent any possible back injuries.
2. Mark the cut you want to make by making a series of light taps into the stone with a striking hammer and stone chisel. Go over the same marks twice more, only harder.

3. Turn the stone over and make the same cuts again on the reverse side, leaning the chisel slightly out into the rock as you strike.

The rock will ideally split exactly where you have made the chisel marks, but in reality, it does not normally happen this way. But, the chiseled cut will usually be enough for you to use it as a building stone. You can continue to work with it until you are happy with the shape.

As you continue to use your stone chisels, you will notice that after hundreds of strikes with the hammer, the tops will begin to mushroom out. It is best to grind these edges off with a grinding stone to prevent any injuries.

Dividing and Splitting Stone

Dividing and splitting stone is handled in much the same process as shaping it:

1. Make several small chisel marks along where you want the stone to split with a chisel tool known as a hand tracer, making sure to hold the chisel tool vertical to the stone.

2. Repeat these same chisel marks at least twice, using more force each time.

3. Turn the stone over and make the same marks on the reverse side of the stone.

4. Use your chisel and hit forcefully in the center of the marked area on the stone. It should begin to split.

PLEASE NOTE: This method is only for sandstone or limestone because these stones are soft and will break relatively easily. Harder stones, such as granite, require different methods, normally involving commercial power tools. Breaking granite is not recommended for beginning stonemasons.

Fieldstones are sedimentary rocks that have formed in layers, which is the reason the rock should break somewhat evenly once you have made the initial splits. Never hit the chisel hard the first time you strike it. It is important that you use the chisel to set a light indention in the stone before you strike hard to prevent breakage in the wrong places. This will also make your work more effective, as well as help prevent any damage to your hands.

Using plugs and feathers (wedges and shims)

There is a widely used technique called the **"plugs and feathers"** method, also called "wedges and shims," although that term also refers to small stone used to fill and level rock structures. This method is used for splitting large stones, and stonemasons have used this practice for centuries. This involves the following steps:

1. Mark the stone with a marker where you want to split the stone.

2. Drill holes about 6 inches apart along the marked split line as deep as your wedges are long. Wedges are thin pieces of metal or wood that you will insert into the hole to cause the split. These wedges have a slightly

curved top and somewhat pointed base. You can make this split either straight or curved.

3. Insert the wedges and shims. You should place shims, which are small metal bits wider at the top than the bottom, into the center and wedges with their "ears," or top curves, pointed outward.

4. Strike the wedges with your stone hammer one by one in sequence until the stone begins to split.

Keep your stones as large as you think you can manage and relative to the size of your project. If you are building an outdoor barbecue, using a stone roughly 2 feet in diameter by 3 feet wide and 1 foot tall is not going to work; cutting a stone that large into several smaller stones would be the wisest choice. However, as a foundation rock for a springhouse or residence, it would be a perfect size.

Carving Stone

Perhaps becoming the next Gian Lorenzo Bernini, a 17th century Italian sculptor, is not on the list of things you would like to achieve in life, but any book about using stone would be remiss in not at least approaching the subject of carving stone.

You can have boulders delivered to your garden and carve them out to create a pool of water for birds and wildlife, or you can use them as an accent in your water garden. You can also carve out bench seats from whole rock, create hot tubs and sinks, and make accents in walls, such as small niches for placing a planter.

Carving stone requires using small power grinders with diamond-chip blades, which create an abundance of dust, making it imperative to always wear dust masks when carving with power tools.

You will **rough-carve** the stone first, which simply means carving the stone in a rough, unfinished manner. As you get your basic shape, **finish-carve** the remainder, which means smoothing the cut and providing more definition. Instructions for carving the projects listed in this book are included here, and all stone carving uses the same techniques, which are as follows:

1. Use a chisel and hammer to score several parallel grooves — also called **kerfs** — in the area you will remove.

2. Chisel out the kerfs using a hammer and large block chisel to break and remove the chunks of stone from the interior. You can use power chisels for this job or finish it by hand using a chisel and hammer.

3. Next, you will begin to finish-carve the grooves, or kerfs, with smaller, more precise chisels until it has reached the desired depth and width.

4. Smooth the surface, which you can do with a power grinder or metal sander.

Mount Rushmore National Memorial is carved into the granite face of Mount Rushmore.

CASE STUDY

Oleg Lobykin, stone carver and sculptor
Stonesculpt
538 Sacramento St.
East Palo Alto, CA 94303
www.customstonecarving.com
oleg@customstonecarving.com
(650) 575-9683

"There is beauty in a stone's polished surface; yet, at the same time, there is beauty in the surface of a rock in its natural state. It is difficult to compete with nature's creations." — Oleg Lobykin

Originally from St. Petersburg, Russia, Lobykin is a stone carver and sculptor, as well as the owner of Stonesculpt, which specializes in custom stone carving and restoration. Lobykin began his career in the United States by working on the Cathedral Church of St. John the Divine in New York City, the largest cathedral in North America. Lobykin was commissioned to do mostly new stone carving but to match specific styles and time periods. He spent six years working on the cathedral with other master stone carvers before moving on to other projects.

Lobykin is also known for carving a column directly into a mountain in northern Alabama, dubbed "The Alabama Column." He worked with the design of a co-artist and friend who was inspired by work she had seen in Egypt and recreated the same style at the face of a quarry located in Russellville, Alabama. "The quarry is all inside the mountain, and it is huge, like a cave. I carved icons that represented the local flavor of Alabama, such as the Confederate flag and red-bellied turtles," Lobykin said.

Lobykin uses all types of natural stone for his work. "It all depends on the design and use for the work. I've worked in all types of marble, including Belgian black marble, granites, limestone of course, and onyx from Mexico. I like to use the onyx because I can incorporate a light source within the stone to make it glow." Many representations of this unique work can be seen on his personal website, **www.lobykin.com**.

Using quarried stone

You can go to your local stone yard and choose stone already shaped into building stones, and this is perfectly acceptable. After all, it is all about the project and not about the creation of your own stone building blocks. It is more cost effective to gather your own fieldstone or reuse old brick gathered from nature.

Commercial quarries break stone using pneumatic and electric drills. Larger jobs are handled using explosives to blow apart the stone. After being transported by huge bucket loaders into trucks specifically designed to carry tons of rock, it is then stacked on pallets in the stone yards and sold per stone or by the ton. You can handpick some stone, but other stone is sold in bulk so you have to take whatever you get.

Now that you have more knowledge about the stones and tools you will need for your projects, the next chapter will talk about the difference between using stone without mortar, called dry stone, and with mortar.

CHAPTER 3: Working with Dry Stone and Mortar

Humans have stacked rocks on top of each other for centuries. Rocks were stacked to create boundaries, cooking pits, and shelter from the elements. Stacking rocks without using an adhesive to keep them in place is known as dry stacking, but through trial and error, people discovered that just stacking stones does not create a durable and long-lasting structure that can withstand the forces of Mother Nature. So, early stonemasons learned to stack rock in way so the structures withstood severe weather and invading animals. They handed down this knowledge through generations, from one farmer to another, from one house builder to another, and from one family to another. Eventually, an adhesive made from sand, lime, and water was invented to help keep these stones firm and solid, and this mixture is what we now know as mortar.

This chapter will explore the techniques of building with both dry-stacked rocks and stone, as well as working with mortar.

Dry Stone

Both terms, dry stack and dry stone, define the same process of building stone walls or structures using interlocking stone and gravity to hold them in place. The term dry stacking refers to the process itself, while dry stone refers to the final product. The classic antiquity of dry stone walls is often associated with the look of the Irish countryside, dotted with walls made from stacking the native stones of the pastures into walls to keep the livestock from wandering into the roads. Experienced dry stone masons in Ireland during the 1700s and 1800s became popular with farmers, and these masons often traveled the country building walls for livestock using all of the stones the farmers dug and plowed from their fields. In return for their labor, the masons received room and board until the stone fence was finished to the farmer's satisfaction. These traveling masons had learned the tricks to keeping walls sturdy and passed this knowledge on to others who have since kept up the trade. The Irish who emigrated from Ireland to America knowing this trade began building walls from unearthed rock and stone to make their living here. Dry stone purists prefer the look of the natural growth of moss and lichen, which tend to grow on dry stone walls, because it creates an appealing landscape accent. There are two methods of creating dry stone walls:

1. Stack the rocks using both large and small rocks to create two walls, side by side, sloping toward each other in the center. You achieve the sloping process by

stacking the stone with the larger ends facing outward, and the smaller ones inward, using a process called **battering**. This method creates a very strong structure with each wall supporting the other. As you build, add larger, flat **tie-stones**, which are stones that span the width of both walls, creating a tie between the two. When you have stacked all your stones, you finish with a capstone, which is a stone that also spans the two walls, creating a solid top surface.

2. The other method also entails building two walls battered toward each other, but the interior gaps are filled with rubble, rock chips, or gravel. This process is called **hearting**. This is the classic form of dry stacking, and it is a quicker method of building the wall and is just as strong. You cannot use small commercial gravel because it will hold dirt and moisture, which during a freeze will push the battering apart.

Mortar

Using mortar, a cement-based adhesive, to keep your stones together is the preferred method of building any structure because of its ease and superior strength. The early Irish working with the dry stone walls did not use mortar for the pasture fencing but did start to use it in the early 1800s for housing because the mortar provided a better insulation than the previously used hay and mud. If you dry stack a wall improperly, it will tumble over time. If you use mortar, it is not likely to come apart on its own. Mortar also seals the stone and makes a more weatherproof structure, retaining heat and keeping moisture out. You can stack smaller stones using mortar, and, if applied properly, it is almost undetectable. You can also use rounded stones without creating flat ends for stacking. A mortared wall can be narrow, while a dry stone wall requires a width of at least two rocks. All mortared structures require digging a foundation or footing to help keep the structure level and provide the strength of a solid foundation. Experts in the field of masonry also recommend digging a footing for dry stone structures, using poured cement or dense rectangular stones as a foundation.

More aesthetically pleasing methods of using mortar exist. One method called **icing** creates an effect where the mortar looks as though it is dripping from the stone. Another method is **roping**, which is smooth ropes of mortar surrounding each stone in a neat fashion. Some mortar is colored with natural minerals or manufactured colorants to bring out textures or natural stone pigments. It is easier to level a mortared wall than to level a dry stone wall because you can make small adjustments in the wall using mortar. Making the mortar joints slightly larger to make the courses level is standard practice.

The following are some basic tenets of mortaring you should implement for any project you choose to do from this book.

Mixing mortar

One of the most important things to remember about mortar is there is a big difference between the drying of mortar and the curing process of mortar. Mortar is a type of cement adhesive, and it, therefore, has the same curing and drying requirements of cement. When water is mixed with cement and sand, the ingredients undergo a chemical change called **curing** that hardens the mortar. If this did not happen, once the water evaporated, you would have cement and sand instead of mortar. When you mix all the ingredients together, curing begins to happen immediately, and it is important to use whatever you have mixed within two hours. You will also need to keep the mortared wall damp as you are working on it so the mortar does not dry before the curing is complete. It is good to have a water hose with a mist attachment handy or at least buckets of water you can use to dampen rags you will use to shield the wall from the hot sun. You can use plastic sheeting to retain moisture on the wall if you need to stop and start again in the morning. If you are working in a cold climate, the chill can also affect the curing process, which is why masons recommend not working in frigid weather. The chemical process of curing takes about two days, and after this time, you should allow the mortar to dry.

There are many time-honored recipes for mixing mortar, but it requires two basic ingredients: sand and cement. There are many kinds of sand and many kinds of cement, and all masons have their own tricks and tips. It helps to know what the different properties of the sand and cement are before you work with them.

Pit-run sand

This is sand used right out of the ground. This is difficult to use, although it is fairly cheap if you live near a beach or other sandy area. It is difficult to use because sand near a coastline contains a large amount of salt and may affect the cement's stability. You will also have pebbles and organic debris if you use sand from around riverbanks or in sandy forest areas. All of these factors will affect how the stones, brick, or rock fit together.

Screened masonry sand

This is the best sand you can use because any dirt or debris is cleaned at the processing plants, and this sand is usually hauled to a job site with a truck. Sharp sand, or sand that feels abrasive when you rub it with your hand, is coarser and will make a stronger mortar, while river silt — also called bottom sand — will make a softer and less durable mortar. Although it is possible to mix sharp sand or masonry sand together with bottom sand, it is best not to because the resulting product will not have predictable qualities of adhesion.

Hydrated lime

Hydrated lime is an additive used in mixing mortar to increase the flexibility and workability of mortar. When hydrated lime is added to mortar mix, the sand and the cement do not separate, and the mortar is more waterproof when dry. Cracking due to mortar shrinkage is generally eliminated when hydrated lime is added to the mixture. Hydrated lime is sold in bags, normally 50 pounds, at all hardware stores that carry Portland cement.

Portland cement

Portland cement makes a hard mortar, which is useful when you need to bond two even surfaces together like in brick structures and when you have wider gaps to fill.

Why is it called Portland cement? James Parker patented "Roman cement" in 1796. James Frost processed "British cement" from Roman cement in 1811, and Portland cement was derived from a combination of "natural" cements made in Britain in the early 19th century. It was named Portland cement after its close similarity to a building stone quarried off the Isle of Portland in Dorset, England. A British bricklayer named Joseph Aspdin patented the process of making a cement in 1824 that became the unofficial Portland cement.

Cement is sold by the bag, which holds 1 cubic foot of cement powder. Cubic feet is how bagged cement is sold. It is also how topsoil for gardens is sold, and here is the formula for converting cubic feet to square feet for measuring purposes:

You have a 12-square-foot garden, and you want to buy topsoil for it. Topsoil is sold by the cubic foot, as is cement, so here is how to determine how much you will need:

The formula for cubic feet is length X width X height.

If you want the topsoil to be 3 inches high, calculate as follows: You know the length is 12 feet, the width is 1 foot, and the height converted into feet is 0.25 feet so you need 3 cubic feet of topsoil.

Masonry cement

This is standard cement used in mortaring uneven surfaces, and it makes a sticky mortar rather than a mortar used specifically for building solid, even structures. It is used for binding more so than strength. It does cure and dry to a hard mortar, as does Portland cement, but it is easier to adhere veneer to a solid surface or two uneven stone structures.

Masons create their own mixtures as they work, depending on what they need to use the mortar for. The general ratio of mixing is one part cement to three parts sand, or 1:3, no matter what mix you use. Adding other elements into the mix, such as lime, gravel, and dyes, will create different results. Here are three alternative mixes for mortar:

- 2:1:9 — This is a mortar mix for general use. It is made from two parts masonry cement, one part Portland cement, and nine parts sand.
- 1:2:4 — This is a mortar mix for even surfaced stone. It is made from one part masonry — or Portland if you are working with concrete block — cement, two parts sand, and four parts gravel.
- 1:2:9 — This is a special mixture for **pointing brick**, a term used when you replace broken or chipped mortar in brick structures to restrengthen the brick and to

beautify it. This mixture also dries to a bright white so keep this in mind when using it. Fill a 5-gallon pail halfway with water and add a bag of **hydrated lime**, which is a powdered mineral used to chemically harden the mortar mixture. Stir until it is the consistency of heavy cream, then leave it to process for a day. Mix a one to nine ratio of Portland cement and sand, and then add one part of the water-lime mixture. A part can be one scoop, one bucket, one gallon — just be consistent, and use the same measuring equivalent for each product.

Mixing mortar is done in a large metal or plastic trough. You can also use your wheelbarrow, which makes moving the mortar to the wall much easier if you are too far from a garden hose. Mix mortar by adding all of your dry ingredients together, adding water to the back of the trough or wheelbarrow. Using your hoe in a back-and-forth motion, chop the dry mix into the water, working it in until you have a dough-like consistency.

Cleaning the mortar

You can clean excess mortar from the joints within a few hours of mortaring by using your pointing trowel and a wire brush. Then, you can sponge off any excess haze with a large grouting sponge and water. If you leave the mortar on for too long and it dries, you will need to scrape or chisel it off. If you have dried stains on the stones, you can use a mixture of one part muriatic acid to ten parts water. Apply this mixture to the stains and flush with water — always wear a dust mask and gloves. Muriatic acid, also known as hydrochloric acid, is a common chemical used in cleaning and etching masonry from rock and stone. It is also used in households for cleaning scale and lime off faucets and showerheads.

NOTE: If you are trying to grow lichens or moss on a mortared fieldstone wall, do not use acid because it will kill any existing mosses or lichens and prevent the growth of more. *More information on how to encourage the growth of lichens and moss on stone will be addressed in Chapter 6.*

Cement mixer.

Mixing Cement

It is also important to note that mortar is not cement, and vice versa. The steps for mixing cement are as follows:

1. To mix cement for a footing or foundation, you will need to mix one part Portland cement, two parts sand, and three parts 1-inch gravel, called six-to-eights in quarries.

2. Always add your water to the dry mix, not dry mix to water, and use a hoe to mix as you add the water. This is important because the dry mix tends to settle on the top of the mixing pan and becomes more difficult to mix.

3. Add your gravel to the water-cement mixture. Adding gravel will strengthen the mix for a more solid and stable foundation. The cement should be dry enough to hold peaks when you shovel it into the trench you have dug for the foundation but wet enough to work with and level out when you place it.

If you have standing puddles of water on your cement mix after you shovel it into the trench, it is too wet and will leave air pockets as the cement cures and dries, which will cause weakness. You should not have air pockets in the cement, and if you do, work them out with the hoe until you find no puddles on the top of the mix.

You must clean all your tools before the cement hardens on them or else you will have a hard time chipping it off. To clean cement from your tools or from other surfaces, you need to wash the tools and surfaces with water after you have finished using them. You can remove some cement from flat surfaces, such as brick, with a wire brush, but it is recommended that you prevent the cement from hardening where it is not wanted rather than cleaning it off after it has set. Once cement sets up and hardens on tools, such as hoes and shovels, it is nearly impossible to get it all off, and the tools are ruined.

After reading about all the tools and techniques of working with stone tools and mixing the cement and mortar, you are ready to choose your project and begin building.

The first project this book describes is how to build is a stone barbecue. Building a barbecue in the backyard is by far the most popular project for beginning stonemasons, and although not the easiest one to build, any novice can complete this task. It uses readily available stone (usually brick), makes use of the mortaring techniques offered in this book, and can be enjoyed by the whole family. The first project involves building one of two types of barbecues — one with a chimney to direct smoke and one without. This next chapter will also explore methods for laying your own patio.

CHAPTER 4: Patios and Outdoor Cooking Pits

Now that you know the basics of choosing your rock and stone, know how to carve and shape stone, and know how to mix your concretes and mortars, it is time to put your knowledge into action. This chapter deals with simple patios, fire pits, and barbecue pits.

Natural Stone Patio

Stone patios are one of the most desired projects for outdoor landscape. You can easily install these patios with either fieldstones cut to the shape you need or with stones you purchase from stone yards. They can be mortared or set into place using gravel and earth. Patios are a good first project and can dramatically improve an outdoor living space. You will need to check local building codes to see if you need a permit to install a patio. You can call your city's building department or ask for the licensing and permitting department and inquire as to whether putting in a patio is considered a building project. Once you have your patio in place, even if the edges are irregular instead of square, trimming sod to fit into the nooks and crannies is very aesthetic and pleasing to the eye.

This patio was built using flagstone rocks.

MATERIALS YOU WILL NEED:

- One and a half to 2 tons of stone, as rectangular as possible — preferably sandstone, flagstone, or brick. Remember: The word ton equates to approximately the amount of stone one standard-size pickup truck's bed can hold. Depending on the size and width of the stones, you could need more or less than 1 ton.
- Six bags of Portland cement, if you choose to mortar.
- One thousand pounds sand.
- Five hundred pounds gravel — crushed stone not river pebbles.

Note: The amount of stone you will need depends on the size of the patio being built. This patio described here is approximately 10 feet by 10 feet. Therefore, take these measurements loosely and adjust according to what area you will need to cover.

TOOLS YOU WILL NEED:

- Garden rake
- Pick
- Shovel with a sharp, flat blade
- Hoe
- Wheelbarrow
- Mason's trowel
- Level

INSTRUCTIONS:

1. Prepare the base for your patio by marking off the area to be laid with stone.

2. Dig out a shallow trench, also called a **footing ditch**, the size of the proposed patio, digging up all grass and weeds and clearing the trench of as much vegetation and rock as possible. This is normally the most time-consuming step. Depending on where you will place the patio and the amount of grass or other organic matter you will need to remove, it could take as long as four hours to properly clean the area for the patio. In more northern climates, it is advised to dig below the frost line, which means adding considerably more gravel to the foundation, around 6 to 8 inches. **Note: You can find your frost line by checking with your local building inspector.**

A patio being prepared for stone placement with footing. Photo courtesy of Douglas R. Brown.

3. Once you have dug the trench, fill it with gravel, leaving 2 inches, or the depth of your stones, of space from the top. To help prevent the growth of weeds, you can treat the footing ditch with a zero-vegetation chemical, such as Ortho® Groundclear® Triox Total. This will kill any vegetation on the site. You can also spread coarse salt on the earth before you spread the stone. Keep in mind that if you want moss or grass to grow between the stones, you should not treat the ground at all.

4. You can also have a concrete slab poured as the foundation but that can be expensive, depending on the size, and if you mix the cement yourself, it will take a week or two instead of a few days to complete your patio.

5. Pack the gravel down as tightly as you can. Walking on it helps settle the gravel. This provides the base for your stones so you want the gravel as tightly packed and even as you can get it. Start laying your stones, keeping them as level as possible. Do not throw another handful of gravel underneath an unlevel stone to level it out; instead, remove the stone and pack more gravel in tightly. In time and with use, the stones will evenly pack the gravel foundation. Use a level to check your progress. Do not lay the stones too close together as you will need to pack earth, crushed gravel, or mortar in between each stone to help hold the patio in place.

6. When you are done laying your stone, you can encourage mosses or grass to grow by filling in the

areas between the stone with peat moss or earth. Of course, this may also encourage the growth of weeds. You can use a commercial weed-killing product to kill the weeds and leave the grass intact. Other methods of filling the spaces between rocks include mortaring the stones and filling the joints with more crushed stone, which is the preferred method because it will settle nicely into the joints, discourage the growth of weeds, and is not subject to cracking, as is mortar, under heavy traffic. When you mortar your stones, rainwater cannot seep between the stones, allowing the water to build up on the stones, making them dangerously slick. Also, mortar tends to break apart under heavy use, and if a stone is not level, it will buckle and crack the joint. Filling the joints with gravel will also allow weeds or grass to grow but are easily pulled out.

Stone, outdoor patio and fireplace.

Barbecue Pit

Barbecue pit that is made out of fieldstones and has a chimney.

One of the most popular outdoor projects is a barbecue pit. This book includes instructions for building two types of pits: one without a chimney for directing smoke and the other with a chimney. The main difference is the pit with the chimney is a little more difficult to build and requires more technique, but even a beginning stonemason can do it with time and patience.

These are instructions for a 36-inch-by-36-inch-by-36-inch, simple barbecue pit with a slab platform on which to build the fire and elevated sides for the grill.

MATERIALS YOU WILL NEED:

- One and a half to 2 tons of stone or brick, as rectangular as you can get it.
- Six bags of Portland cement.
- One thousand pounds sand.
- Five hundred pounds gravel.
- One 50 pound bag of lime.
- One 24-inch-by-30-inch-by-5/8-inch sheet of plywood, cut to 24 inches by 24 inches.
- Five two-by-four wood boards each 36 inches long.
- Five rebar pieces cut to 26 inches long and five pieces cut to 24 inches long.
- Cable ties.
- A box of #10 nails.
- Metal grid, at least 24 inches wide.

TOOLS YOU WILL NEED:

- Pick
- Shovel
- Hoe
- Wheelbarrow
- Trowel
- Pointing trowel
- Wire brush
- Hammer
- Gloves

INSTRUCTIONS:

1. Dig a foundation trench, also called a footing ditch),in a square U-shape 12 inches wide by 42 inches long

on each side. The foundation trench should be about 6 inches deep or below the frost line. **Note: You can find your frost line by checking with your local building inspector.**

2. Mix your cement in a wheelbarrow with enough water for the mixture to form a workable consistency. Start with two shovels of Portland cement, four shovels of sand, and six shovels of gravel.

3. Shovel this mixture into the trench, and continue mixing cement and filling the trench. This process is called pouring the footing, or foundation. You do not want to stop and start this step of the project, as it is important to get all the concrete mixed and poured within a two-hour time frame.

4. Dampen the mixture with a misting attachment on your garden hose, and cover it with a plastic tarp for two days to allow the cement to cure. Remember: Curing is the process in which the chemical reactions take place to harden the cement. Drying takes place after the curing process has occurred. If the cement dries out before the curing process is complete, the concrete's strength will be compromised, and it will crack.

5. When you are ready to mix the mortar and start laying the stone or brick, begin with nine shovels of sand, two shovels of cement, and one shovel of lime, plus water.

6. Lay your stones, centering each course in the middle of the foundation, to reach a height of approximately 30

inches. The courses should be one stone wide. **Note: Courses are the single length of stones within a wall, which are usually two courses thick.**

7. When your sides are built to the height of approximately 30 inches, you will begin the rebar grid that will reinforce the slab on which the fire is built. **Rebar** is a popular abbreviation for "reinforcing bar," which is a carbon steel bar formed with ridges for anchoring concrete. Using rebar gives strength to any concrete structure. Lay the longer lengths of rebar across the sides of the stone walls, and then overlay the shorter pieces across. Secure the overlaps with cable ties, and cut the loose ends off the cable ties to prevent them from sticking out of the poured cement. Set the grid far enough back on the rear wall to avoid having any rebar jutting out from the cement in the front.

8. Position the 24-inch-by-24-inch piece of plywood about 2 inches under the rebar grid, nailing four of the two-by-fours into the plywood as legs to support the platform. Use the last two-by-four as a lip nailed to the front of the plywood and butting up to the rebar with approximately 1 inch of clearance.

9. Add another 6 inches of stone on top of the side and rear walls.

10. Mix more cement and pour over the plywood form, encasing the rebar with at least 4 inches of cement.

Trowel it smooth, and cover with the plastic tarp, keeping it moist for three to four days.

11. Remove all the wood bracing by tapping with a hammer until it loosens from the concrete. Backfill around the base of the pit with soil.

12. Before you use the outside grill, pour sand on the top of the slab to protect the concrete from direct heat to prevent possible cracking and discoloration. Lay your metal grid on the top of your side and rear walls.

Barbecue Pit with Chimney

The stone fire pit and chimney above is constructed of slate.

This stone fire pit and chimney is made out of fieldstone.

By building a barbecue pit with a chimney, you can control the smoke that results from grilling. Aesthetically, this barbecue pit looks more appealing in your landscape than a barbecue pit without a chimney.

MATERIALS YOU WILL NEED:

- Three to 4 tons of stone or brick, as rectangular as you can get it.
- Eight bags of Portland cement.
- One thousand-five hundred pounds sand.
- Seven-hundred-fifty pounds gravel.
- Two 50-pound bags of lime.
- One 24-inch-by-48-inch-by-5/8 inch sheet of plywood, cut to 24 inch by 24 inch.
- Five two-by-fours, each 36 inches long.
- Five pieces of rebar cut to 26 inches long, and five cut to 24 inches long.
- Cable ties.
- One box of #10 nails.
- Metal grid, at least 26 inches wide.

TOOLS YOU WILL NEED:

- Pick
- Shovel
- Hoe
- Wheelbarrow
- Trowel
- Pointing trowel
- Wire brush
- Hammer
- Circular saw
- Gloves

INSTRUCTIONS:

1. Dig a foundation trench, also called a footing ditch, in a square U-shape 12 inches wide by 60 inches long on each side. The foundation trench should be about 6 inches deep or below the frost line. **Note: You can find your frost line by checking with your local building inspector.**

2. Mix your cement in a wheelbarrow with enough water for the mixture to form a workable consistency. Start with two shovels of Portland cement, four shovels of sand, and six shovels of gravel. Shovel this mixture into the trench, and continue mixing cement and filling the trench. Dampen with a misting attachment on your garden hose, and cover with a plastic tarp for two days to allow the cement to cure.

3. When you are ready to mix the mortar and start laying the stone or brick, begin with nine shovels of sand, two of cement, and one of lime, plus water. Lay your stones, centering them on the footer, to reach a height

of approximately 30 inches. The walls should be one stone wide.

4. When your sides are built, you will begin the rebar grid, which reinforce the slab on which the fire is built. Lay the longer lengths of rebar across the sides of the stone walls, and then overlay the shorter pieces across. Secure at the overlaps with cable ties, and cut the loose ends. Set the grid far enough back on the rear wall to avoid having any rebar jutting out from the cement in the front.

5. Position the 24-inch-by-24-inch piece of plywood about 2 inches under the rebar grid, nailing four of the two-by-fours into the plywood as legs to support the platform. Use the last two-by-four as a lip, nailed to the front of the plywood and butting up to the rebar with approximately 1 inch of clearance.

6. Add another 6 inches of stone on top of the side and rear walls.

7. Mix more cement, and pour over the plywood form, encasing the rebar with at least 4 inches of cement. Trowel it smooth, and cover with the plastic tarp, keeping it moist for three to four days.

8. Remove all the wood bracing by tapping with a hammer until it loosens from the concrete.

9. Prepare a plywood form for the arch that will form the base of your chimney. Make it 24 inches wide

and 8 inches tall. With the 8 inches in the center, draw a curve from the center to each end, dipping down to 6 inches on each side. Cut two of these same forms with the circular saw. Cut four blocks of wood, approximately 4 inches wide. These will be the spacers in between the plywood pieces. Nail in place, creating an arched form.

10. Place your form on the ground and dry-fit — selecting the stones that will fit together well to create the arch — allowing for ½ inch of mortar in between each stone. When you have all your stones selected, place the arch on two wooden wedges on the center of the stone slab, 30 inches from the rear wall.

11. Begin laying your arch stones, starting with pairs of tapered stones at each end of the form. Keep all your mortared edges pointing toward the center of the arched form. Continue mortaring your stones from each end, setting the keystone, or center stone, last. Let the arch set up for a day before removing the form.

12. Blend the arch stones with the wall stones by filling in any irregularity to form a horizontal, level surface. Begin stepping in the side walls 2 inches at a time. You will also need to step in the front of the chimney, but the back should remain parallel with the rear wall.

13. Build the chimney (four walls) as high as you want it, but an optimum height should be no more than 6 to 8 feet.

14. Back fill around the base of the pit with soil.

15. Before you use the outside grill, pour sand on the top of the slab to protect the concrete from direct heat. Lay your metal grid on the top of your side and rear walls. You will also want to burn crumpled paper in the back of the chimney to help get the heated air from the fire on the slab to rise.

Fire Pit

You can use fire pits for the same thing you use barbecue pits for, although they are used more for ambience than for cooking. You use fire pits with an open flame, while you use barbecue pit with an enclosed flame. You can also use the fire pit as a more elaborate and safe form of campfire to help you keep warm during family gatherings in the backyard.

MATERIALS YOU WILL NEED:

- One to 2 tons of flat, rectangular stone.
- One-quarter ton flagstone for capping the fire pit.
- Half to 1 ton ¾-inch drainage gravel or pea gravel.
- Masonry adhesive for caulking gun.
- Iron campfire ring.

TOOLS YOU WILL NEED:

- Pick
- Shovel
- Wheelbarrow
- Gloves
- Stone chisel
- Mason's hammer
- Level
- Caulking gun
- Rubber mallet

INSTRUCTIONS:

1. Lay a ring of blocks around the area you want to put your fire pit, placing them end to end until you have your circle. You may need to cut the blocks to fit one or two of them into place, in which case you would cut the stone using the chisel and mason's hammer. Place the block to be cut over the area you need to cover, mark with a pencil, and score with your chisel. Cut a

deeper line with the chisel, and hit with the hammer until the block splits.

2. Mark the location for the pit by digging around the block ring with a shovel, approximately 1 inch deep.
3. Create a level trench foundation for the pit by digging a straight-sided trench 12 inches deep and as wide as one block. You will insert a ring made from individual stones into this trench. Dig down an additional 6 inches inside the block ring, and remove any stones you used for measurements.
4. Fill the trench with 6 inches of ¼-inch drainage gravel or pea gravel. Compact this as much as you can by stepping on the gravel to condense it down into the ring. Keep the ring level.
5. Lay and level the first course by placing the first block in the outer ring filled with gravel. Check to make sure this block sits level by using a 2-foot level. If the block is too high, tap it with a rubber mallet until it sits level. If it is sitting too low, place a bit more gravel underneath it, and tamp it down.
6. Continue to lay blocks around each other, and finish the ring. The blocks need to butt up against each other tightly. If the last block does not fit, try to fit it into place using the mallet. If it is too loose, the fire pit will not be effective, and you may lose heat or sparks

could set the surrounding vegetation on fire. Check to make sure the stones are level as you lay the blocks.

7. Using a caulking gun, squeeze a double bead of masonry adhesive across two adjacent blocks. Lay a block on top of the adhesive-covered blocks, centering it over the joint between the two bottom stones. Continue until the second course is finished.

8. Fill the pit with approximately 6 inches of gravel, which will help support the first two courses as they set up. Lay the third and fourth courses using the adhesive, centering over the bottom joints as you go. Place the iron campfire ring into the circle, and adjust it using the mallet to sit level with the top of the block wall. Fill in any spaces with gravel.

9. Place the flagstone cap pieces on top of the pit walls. There should be approximately 2 inches of overhang off the top of the block wall. Try to fit the stone tightly together, cutting the stone as needed. Adhere these top stones to the block by using mortar or masonry adhesive. Wait two days before using the fire pit.

Wood-fired Kiln

These are beehive-shaped Wildrose Charcoal Kilns. In the late 1800s, they were used to produce charcoal for ore smelters in the Panamint Range at the western edge of Death Valley National Park.

Wood-fire kilns, although not complicated in theory, are a bit difficult to use and can be dangerous if not watched properly. They throw off an enormous amount of heat and spark, and when firing some forms of pottery, they require hours of feeding wood to it to maintain a certain temperature for setting the glaze. Local zoning laws also may not permit the use of a wood-fire kiln on your property so check with your local city offices for more information.

MATERIALS YOU WILL NEED:

- Half ton of solid red brick or firebrick, which is specifically manufactured for the high heat in fireplaces.

- Quarter ton flagstone for making the entrance area to the kiln.
- Steel or iron grate, approximately 16 inches by 16 inches.
- One iron sheet, approximately 14 inches by 14 inches.
- Iron strips (scrap).
- Wood for fuel.

TOOLS YOU WILL NEED:

- Shovel
- Rake
- Wheelbarrow
- Gloves
- Level

INSTRUCTIONS:

1. Select a site for your small kiln at least 20 feet from any trees or structures. Level the site using a shovel and rake, ensuring no grass or plants remain where you want to place your kiln.

2. Lay a square of bricks directly on the surface of the soil, leaving one side open. If you are using a 16-inch-by-16-inch metal grate, your bricks should be 14 inches by 14 inches, which will allow the metal grate to rest on top of the second course of bricks. Use a level to make your surfaces even.

3. Lay two more courses of bricks on top of the metal grate. Insert your iron sheet on the top of the fourth course of bricks, laying it across in a diamond fashion.

The corners will face the front, back, and sides of the partial square formed by the bricks.

4. Place some scraps of iron sheeting along the top level of bricks to partially cover them, creating a level surface. Begin to add more brick directly on the iron sheet in the shape of a rough circle. This creates the chimney for the heat to escape.

5. Keep adding brick in a circular fashion, with each course slanting inward and forming a smaller circle toward the top. Wherever a hole has formed from adding the courses, place a piece of scrap iron. Leave an opening roughly 6 to 8 inches in the top.

You will add firewood onto the grate to fire your kiln, with your pottery sitting on the iron sheet. Always keep an eye on any wood fire, and never leave it unattended until the kiln is cool and not throwing any spark. Always leave the pottery in the kiln until it is somewhat cool to the touch.

Now that you have mastered the techniques of building a patio and barbecue grill, you can complete more projects that will add variety to your outdoor landscape. Building basic stone planters and benches will make use of your skills in using tools as leverage and introduce you to using power tools made for carving stone.

CHAPTER 5:
Enhancing Your Landscape

All of the outdoor uses of stone — from building a patio or a barbecue pit to adding stone steps or pathways in your garden — do as much or more than having a professional landscaper come in with bags of mulch and plants. Most of these outdoor accents are easy for the first-time stonemason and are relatively inexpensive.

Basic Stone Planter

Stone planters can be a huge boulder carved out to form a basin in which to put plantings or rectangular stones mortared together to form a long, boxy shape. For the beginner, building a stone planter with rectangular brick or stone is the easiest project to take on, but one large stone carved out and placed in a prominent position in your garden is quite impressive.

MATERIALS YOU WILL NEED:

- One to 2 tons of stone or brick, as rectangular as possible.
- One 50-pound bag of Portland cement.
- Two hundred pounds sand.
- One 50-pound bag of lime.
- Twenty pounds gravel.

TOOLS YOU WILL NEED:

- Shovel
- Hoe
- Wheelbarrow
- Trowel
- Pointing trowel
- Wire brush
- Gloves

INSTRUCTIONS:

1. Begin by planning the length and width of your planter. To eliminate the need to break your stone or brick, try to make it work out evenly by planning the width first and the length second.
2. Dig a trench, also called a footing ditch, as wide and as long as you want your planter. Four inches should be deep enough for the foundation, but you may want to dig below the frost line (about 6 inches). **Note: You can find your frost line by checking with your local building inspector.**
3. Fill with crushed gravel, within 2 inches of the top.

4. Mix the mortar using nine shovels of sand, two of cement, and one of lime, plus water.

5. Begin laying your brick or stone, creating a base of one brick, mortared with other bricks on the sides. You will need ¼ inch to ½ inch of mortar between each stone but no more.

6. Begin creating the second level by overlapping the brick or stone where the joints are on the first level. This gives the planter strength and prevents cracking at the mortared seams. Continue until you have reached the desired height.

7. As always, allow to cure for approximately two days and then dry. Fill with earth and plant.

Carved Stone Planter

If you have a large boulder of any type of rock already in place in your landscape, you can easily create a planter with a cutting blade. Depending on the region of the country you live in, it may be made of limestone, granite, or sandstone. If you do not have a large boulder in your existing landscape, you can have a local stone yard deliver one. Make sure you know exactly where you want the boulder so you do not have to move it once it is in place, before or after you have shaped and carved it.

MATERIALS YOU WILL NEED:

- One boulder — this rock can be as big as 1 or 2 tons or as small as ½ ton.

TOOLS YOU WILL NEED:

- Cutting blade and air compressor
- Stone grinder attachment
- Sledgehammer
- Mason's hammer
- Large and small chisels
- Gloves
- Earplugs
- Goggles
- Charcoal pencil

INSTRUCTIONS:

1. Begin by planning how wide and deep you want the basin. Using a charcoal pencil, sketch the rounded shape on the flattest part of the rock. It is best if the carved basin within the stone is at least 12 inches wide.

2. Using your stone chisel and hammer or your mason's hammer, score the center of the outline as deeply as you want the planter. **Scoring** means creating parallel lines in the rock that will ease the carving process by cracking the surface of the rock. Begin scoring outward just as deeply as the center mark. If you want to use the boulder as a water basin for birds or other wildlife, you need to keep the basin shallow — no more than 3 inches deep.

3. When you have completed the initial scoring, begin to cross-cut with the cutting blade, creating small squares of rock. This should allow you to then chisel out these small squares until you have a quasi-circular shape within the stone. Once you have used the grinding tool to fully round out the basin, you will begin to see the circle take shape.

4. Keeping a firm grip on the grinding tool, grind the interior of the basin as smoothly as possible. Remember to wear goggles, a dust mask, and earplugs because you will generate a significant amount of stone dust and noise. Keep your garden hose handy to cut down the stone dust residue and to cool your grinding blade.

Stone Benches

Essentially, a simple stone seat is nothing more than a large, horizontal slab of stone set on two other large, vertical stones, set at angles to prevent tipping. The slight angle keeps the top slab in place using counterbalance, and as the stones settle into place, the seat naturally becomes more level. It only takes three stones to make a bench, and you will not need cement or mortar. The biggest issues you will encounter when making a bench will be the weight of the sitting stone and moving it into place.

MATERIALS YOU WILL NEED:

- One large, smooth stone slab, approximately 18 inches by 36 inches and at least 4 inches thick.
- Two additional stones for legs, at least 4 to 6 inches thick and 18 inches by 18 inches long, each with one straight edge. You may have to chisel the edges.

TOOLS YOU WILL NEED:

- Level
- Stone chisel
- Hammer

INSTRUCTIONS:

1. First dig a 6- to 8-inch deep trench, also called a footing ditch, where the leg stones will sit approximately 24 inches apart from each other.

2. Set each leg stone in the foundation, and secure by backfilling.

3. Set the large stone on top of the two leg stones. You will need help to do this because a slab large enough for a stone seat will be very heavy.

4. If necessary, you may have to use small stone shims or wedges to make the sitting slab level. Try to choose the smoothest side of the slab so it does not have too many rough edges or pockets that will hold water or be uncomfortable to sit on.

CASE STUDY

Pablo Solomon, artist, designer, and stone sculptor
musee-solomon
7075 W. FM 580
Lampasas, Texas 76550
www.pablosolomon.com
musee-solomon@earthlink.net
(512) 564-1012

"Stones have spirits. If you listen, they tell you what wonders they have seen and what lessons they have learned. The stone has waited countless centuries to show the world the beauty hidden within. Like a surfer riding a wave, I never fight the stone. I allow the flow of the stone to guide me. I never remove any more material than the stone wants me to." — Pablo Solomon

Pablo Solomon is an internationally recognized artist, renowned for his fine art in sketching, mixed media, and design. It wasn't until he turned 50 that he took on the challenges of working with stone.

"I was right at 50 when I decided to try sculpting. Beverly (his wife, muse, and creative director) and I had just visited Italy and seen Michelangelo's series of sculptures in Florence, which he started but never finished. They also had a display of his tools. Because his sculptures were in various stages of completion, I could see what he did and how," Solomon said. "When I returned home, I did a relief sculpture in limestone. Just as I had finished it, a visitor to my gallery (musee-solomon) bought it on the spot."

Solomon still prefers to work with the same tools people have used for thousands of years: a hammer, a pointing tool, a chisel, and a smoothing chisel. "The only machine tools I use are a power drill and an angle grinder with a rock cutting blade as these save a lot of rough-out time."

Solomon prefers to work with a local limestone from Texas, a hard, durable stone that takes a great polish; yet, it is not as hard as granite or marble. He also uses a local quartzite/sandstone because it finishes out to a beautiful grained mahogany. "I've also done work in other local stones, among them serpentine and marble."

Solomon has restored stone walls and rebuilt some entire stone buildings on his 43-acre property he calls his "piece of paradise."

When asked about the best words of advice he could give beginning stone sculptors, Solomon said, "The biggest mistake that I made was over doing it. You really must pay careful attention to the stress you are putting on your body. Also, it is important to know stone. Study the amazing skill of the best stone workers. I am still in awe at what a skilled stonemason can do."

Stone Steps

Building simple stone steps, such as mortared porch steps or terraced steps used in landscaping, are attractive and sturdy. **Mortared steps** are simply rows of stone or brick held in place with mortar, while **terraced steps** are held in place by earth. When making mortared steps, you need to remember that the base step will need to be the largest step so it can provide a surface for the final step. In terraced steps, the steps are inserted into the ground, and they use the natural slope of the ground to provide the height.

Close-up of granite steps.

Mortared steps (three steps, 48 inches wide)

MATERIALS YOU WILL NEED:

- One ton of rectangular stone or brick, at least 6 to 7 inches wide.
- Four 50-pound bags of Portland cement.
- Seven hundred pounds of sand.

- Five hundred pounds of gravel.
- One 50-pound bag of hydrated lime.

TOOLS YOU WILL NEED:

- Pick
- Shovel
- Hoe
- Trowel
- Pointing tool or trowel
- Wheelbarrow
- Level
- Stone chisel
- Hammer
- Gloves

Note: All of the above measurements are approximate as to how many stones you will need. These instructions are basic. The riser height, the face of each step, is determined by how many steps you need. Risers generally should be no more than 6 to 7 inches.

INSTRUCTIONS:

1. Dig a rectangular-shaped trench 48 inches out from the patio, porch, or other area you are building the steps for. The trench should be approximately 6 inches deep or below the frost line in areas with cold climates. **Note: You can find your frost line by checking with your local building inspector.**
2. You will need approximately six wheelbarrow loads of cement for the footing. Mix using the standard recipe: two shovels of cement, four of sand, six of gravel, and water.

3. Mix your dry ingredients together, with the exception of the gravel, in a mortar pan. Add water until the mixture is the consistency of thick pancake batter, and then add the gravel and mix again.

4. Pour the mixture into the footing ditch to ground level. Allow the cement to cure for at least two days. **Note: You can also use concrete block instead of pouring a footing ditch to speed the process along.**

5. Lay the first step using flat-sided stones, sloping them slightly to the front as you lay them to prevent water from settling on the steps. Lay the stones, allowing for no more than ¼-inch wide mortar joints. Lay a complete box of stone, approximately 48 inches wide.

6. Lay the next step on top of the box of stone you created, overlapping the first step by 1 or 2 inches.

7. Repeat this procedure for the third step.

Check your building codes by calling your city licensing and permitting department because in some states, porches that have a height of 30 inches or more must have a railing. Building codes for each state can also be located by accessing Reed Construction Data online at **www.reedconstructiondata.com/building-codes**. You can order a metal railing or build one yourself from wood. You can find easy, free instructions on simple porch railings at eHow.com (**www.ehow.com**).

Terraced steps

Stones used for terracing are normally large, naturally flat stones, such as thick-cut flagstones. They must be supported by earth, not by concrete or mortar, and should not be thin, or they will crack and split. Your stones will need to be 4 to 6 inches thick. If you do not live in a hilly area, you must have enough soil delivered to create a berm for terracing these steps. A landscape **berm** is mounded soil, typically 1 to 3 feet tall and approximately four times as long as its height. It is added to flat landscapes to add interest and dimension. When you have a berm added to your landscape, you can add terraced steps to the back or front walkways, even if you do not live in a hilly area.

Terraced limestone steps.

MATERIALS YOU WILL NEED:

- One ton of flat, square, or rectangular stone — flagstone — at least 24 to 36 inches long and wide. Each of these stones will become one singular terraced step on the hill or berm.

TOOLS YOU WILL NEED:

- Pick
- Shovel
- Hand spade
- Gloves

These steps are the easiest to build because all the project requires is digging earth and inserting stones. However, it is labor intense so you should prepare for the amount of time and effort you will put into this project. Digging into a hill of earth requires digging into the soil and inserting the stones, placing the earth back as a support for each stone. It can take weeks to properly build terraced steps, but the effort will be worth it when you see how lovely your garden area looks with the added interest of a natural walkway.

INSTRUCTIONS:

1. Dig an insertion trench to place the stones in at least 12 inches into the mounded earth.
2. Insert the first stone step, packing in the earth and bracing the step by tightly packing the earth underneath the stone. Try to make at least 16 inches of step available to walk on. Anything less than a 16-inch tread is hard to negotiate for most people when climbing up or down steps. You must pack the earth tightly underneath each step so the steps are solid.
3. Dig another insertion trench for the second stone, about 2 to 3 inches above the first stone, following the natural slope of the terrace. Again, pack with earth.

4. Continue using the stones to overlap one another until the terraced steps are complete.

Using Stone as Garden and Pond Edging

Stones add an additional natural element to your landscaping, and using stones as edging helps keep things tidy. Edging contains your landscaping and provides a visual border from the lawn to the plantings. Bordering flower beds, pathways, and ponds provide the simplest projects you can try using fieldstone or chiseled stone. Any stone will work, but choose stone colors and shapes that enhance your landscape and complement your home. For example, if you live in Vermont, natural granite and greenstone perfectly complement garden areas, and these types of stones are readily available at local stone yards. If your home is built from sandstone, use sandstone pavers to edge your landscaped areas.

Field stones were used to create a small retaining wall around this pine tree.

Here recycled concrete was dry stacked to create a small wall. Photo courtesy of Douglas R. Brown.

A stone mason is creating a retaining wall at a winery in Sonoma, California using fieldstone. Photo courtesy of Douglas R. Brown.

Stone edging

MATERIALS YOU WILL NEED:

- Several stones, different shapes and sizes, or use cut stones or brick for a more formal look. Again, you will need to measure the area you need stone for, and, using a standard of each stone being approximately

6 to 8 inches long, estimate the amount of stone you will need. For example, if you have measured your garden edging at 260 inches, divide 6 inches into that figure. You will need approximately 44 stones to complete this area.

QUICK TIP: To measure a round or circular garden area, use a 75-foot garden hose to lay out the dimensions of where you want your edging. As you set the hose, drive small stakes in to indicate where you need to lay the edging. Using a hose as opposed to using mason string will allow you to bend around the landscaping easily and not have to tie string to separate posts. You can move the hose around and change the dimensions without pulling stakes.

TOOLS YOU WILL NEED:

- Shovel
- Hand spade
- Gloves
- Garden hose, at least 75 feet
- Fifty to 100 small garden stakes

INSTRUCTIONS:

1. Lay out the garden hose in a circular fashion around the garden areas. Drive small stakes in the ground where the garden hose is, and continue to do this, moving the hose as you go, until you establish the entire boundary of your garden edging.

2. Dig a trench around the planted landscaping, removing weeds or grass and making it about 2 to 3 inches wide and approximately 2 inches deep.

3. Begin digging into the trench and setting the rock. Pack the earth around the front and back of the stone. Lay another rock as close as possible to the first one. As they age and weather, the earth around the rock will encourage the growth of moss and lichen.
4. Wet the entire edging with a garden hose to settle the earth.

These steps are for a single rock-height edging. For a higher edging, you will need to use mortar to place additional stone on top of the single stone edging.

To mimic a natural pond edging, you will need to place larger rocks around the direct edge and gradually decrease the size toward the outside of the pond itself. You can add spaces to leave room for native plants that will begin to grow and create a more natural landscape. .

Paths

A stepping stone pathway.

Slate pathway.

A path with stepping stones is relatively easy to install and complements any landscape pattern. This path can easily become a permanent part of your landscape, and the amount of rock you will need depends on the length and width of your path. Use the following directions to make a path that is 3 feet wide by 35 feet long.

MATERIALS YOU WILL NEED:

- One ton of fieldstone.
- One ton of irregular flagstones, at least 21 inches thick and 18 inches wide.
- One ton of very small gravel, also called pea gravel, because the stones are usually no larger than the size of a small green pea.

TOOLS YOU WILL NEED:

- Shovel
- Wheelbarrow
- Trowel
- Stone chisel
- Hammer
- Gloves

INSTRUCTIONS:

1. Plan out the pathway's positioning, and remember pathways that meander around curves are always more aesthetically appealing than straight paths. Plan for the path to curve around trees, stone groupings, or garden plantings. You can make a quick sketch of your yard, drawing it from above, and this will help you visualize where the pathway should go, or you can drive stakes into the ground where you want the pathway to go.

2. Dig a trench approximately 2 to 3 inches deep for the path's edging. As you go along, even out any rough places in the landscaping. Set your border stones, fieldstones, along the exterior edges of the trench, pushing in soil around the stones to set them.

3. Dig out a level base for the interior stones that will make up the actual walkway itself. This should be approximately the width of the flagstone, plus 1 inch for the base of pea gravel. Fill in the trench with the pea gravel to about 1 inch below the ground level. Lay your stones in place on the gravel, leaving approximately 2

inches between each stone. Add additional gravel in between the stones to level the height. Using a garden hose, soak the pathway to help the soil and gravel settle. Adding water will help compact the gravel and soil naturally. Leave the path unused until it dries, which will take about two days. As the stones settle even more with use, which should take about a week, add more gravel to fill in any sunken areas.

The most common wall used in landscaping is known as a dry stone wall and involves stacking stones in such a manner that they remain sturdy and solid. Although it takes years to master the professional techniques of dry stacking, this book will give you the instructions on how to build a simple dry stone wall that will add beauty and privacy to your landscape. This wall, when built properly, will last for hundreds of years.

CHAPTER 6:

Building Stone Walls

Sandstone wall.

Almost every stone structure is a wall of some sort. When you build a barbecue, it has four sides called walls. Laying one stone on top of the other when building a wall is called **creating a course**. Most stone walls are made from two courses laid side by side. Once you have constructed any project where you lay two stones on top of each other, you have succeeded in learning the basics of a wall.

Tie-stones and Capstones

The gravity and slope of the two, or more, stones together will not prevent the wall from toppling over. As you build the wall, the stones will gradually begin to level. To keep the wall sturdy, you will need to use tie-stones. **Tie-stones** are used for structural strength only, but because you can see them in the wall, use the best stones you can for this. Larger, flat stones will make the best tie-stones for your projects. As you continue to build the wall, you will overlap these stones every 1 to 2 feet. Because the tie-stone needs enough width to span your wall, set any possible tie-stones apart from the other fieldstones so you have plenty of tie-stones to work with. **Capstones**, which are large, heavy, flat stones used to span the entire width of the wall at the top, need the same properties of the tie-stones. Capstones are heavier than the tie-stones and help to weight the wall into place.

Shaping Stones

Not every stone comes prepared for a dry stone wall so you may have to square your stones up or split them to make best use of your supplies. Round stones will not make a good dry stone wall, unless you use mortar to keep them in place. Therefore making your wall a mortared wall and not a dry stone wall. Even then, it is very difficult to mortar round stones and maintain structural integrity.

To shape your stones, set the stones on a flat surface and use your stonemason's hammer to hammer a scoring mark along the edges of the stone you want to remove. Remember to wear your goggles to avoid any possible damage to your eyes by flying pieces of rock. You want your building stones to be somewhat rectangular and flat in shape, like bricks. *This will mean that you shape the stone with your chisels and stonemason's hammer to create a rectangular shape, as mentioned in Chapter 2.* This allows you to stack the stones easily. When you start stacking, you will notice stones do not always fall in to place, and pockets of empty spaces occur. You will fill these in as you go with **riprap** — a mix of broken rock, gravel, and other crushed rock debris — and small stones created from shaping the larger ones. Nothing goes to waste when you build a dry stone wall.

As you build your wall, you will want to cover all your vertical joints with another stone and always overlap the joints as you go to keep all your stones interlocked. To make your wall stronger, use all the long stones you can find as interlocking stones. As you work, remember to stand back and look at your progress. You can avoid many potential problems, such as forgetting to cover joints or to keep the wall battered, just by looking at what you have already done.

Using Shims and Hearting

Shims are stone wedges that help fill a space and support the upper and lower stones. They are used fairly often in dry stone walls. **Hearting**, which is comprised of riprap, other rock debris, and gravel, is also used to fill the empty center spaces to help provide stability and level surfaces for the stones to sit upon. Keeping your wall level as you build up is crucial, and adjusting your levels to accommodate the natural slope of the ground is important.

Finishing your wall and adding capstones

Many dry stone walls are finished using small, flat, vertical stones, as in the dry stone walls of Ireland and Kentucky. Individuals used this practice to keep people from sitting or walking on the walls. Other walls are finished using large, flat capstones across the width of the wall, as near thickness to each other as possible. You can purchase quarried fieldstone or sandstone already cut square and flat to finish the top of the wall if you choose, and it adds a clean finishing touch.

Building a Corner

If you want to add another wall, you will need to build an interlocking corner, which helps strengthen both walls. Even if you choose to build another wall at a later date because you are dry stacking the wall and not using mortar, you can disassemble the wall to add the corners at your leisure.

Begin building your second wall at the corner of the first wall, keeping the stones of the base close to the end of the adjoining wall. Begin overlapping your wall stones and your tie-stones. Create your corner by overlapping capstones, creating a vertical wall end. Ultimately, the corner will be stronger than the straight wall.

There are two basic types of walls: mortared walls and dry stone walls. *Chapter 3 identified the differences between the two.*

CASE STUDY

Chuck Miller
Vice president of Paverscape, Inc.
St. Cloud, Florida

Before building a retaining wall, you should thoroughly plan the project and carefully examine the land. "The most important aspects of building a wall are proper engineering and understanding the engineering methodologies associated with segmented retaining walls," said Chuck Miller, vice president of Paverscape, Inc.

Understanding the requirements for a successful wall and following instructions will help prevent structural problems in the future. One of the most common problems homeowners experience when building a retaining wall is having soil on site not suitable for a wall. Homeowners should remove the bad soil and replace it with good soil.

"Soils that are considered unsuitable are clayish or putty-like. The best soil for retaining walls is one that is sandy in nature. Each sand particle has jagged edges that create friction when gravity and pressure are applied. Water will flow through sand much easier than clay," Miller said. This flow of water will prevent a buildup of water behind the wall, which could cause the wall to crumble.

When building a wall on your own, it is imperative that you properly complete all steps in sequence. "Every component and process of a retaining wall is equally important and vital when designing and constructing because it is a systems design. This means that every single item and process supports the others, and if you eliminate or dismiss one of them, it will have a negative effect on the structural abilities of the wall," Miller said.

"There are many applications for a retaining wall, but the overall common use is when there is an elevation change on a property and the homeowner wants to make a distinction in that change. For example, a retaining wall positioned behind a pool deck adds a nice transition to the backyard. Some additional examples include tree rings, seating walls, planters, and elevated patios," he said.

The average person can easily build smaller projects, but larger walls — more than 3 feet tall — that need engineering will require help. "Most retaining walls should be built by a qualified and specialized retaining wall contractor who has engineering capability," Miller said.

Building a Dry Stone Wall

There are two kinds of dry stone walls: those created as walls for defining property lines and to harbor livestock and those created as retaining walls to hold back an embankment and create a terraced look to your landscape. Both are created somewhat the same, and both are referred to as dry stone walls.

Freestanding dry stone walls

A dry stone wall at Batsford Church in Gloucestershire, England.

Harnessing the principles of gravity and using interlocking stones are the key elements in building a freestanding dry stone wall. Without these two elements, your wall will tumble, leaving you with a pile of rocks instead of a wall. If you initially learn to create a dry stone wall, creating mortared stone walls becomes much easier. The classic dry stone wall has stones that are seen from both sides with chunks of gravel or broken rock wedged in between the courses for stability, called hearting. The gravel is not seen from the outside. A dry stone wall is constructed by using a technique called **battering**, which entails creating two walls that stand together and lean in toward each other, becoming slightly narrower at the top than the base.

PLAN THE LOCATION

You need to draw on paper where in your landscape the wall will start and finish, referred to as "laying the groundwork." You will need to mark these areas by pounding in stakes and running a string line from stake to another, laying the base course for the wall. Measure these areas and make note of the length of the walls to determine the amount of stone you will need to complete the project. Planning the location and how much stone you will need, as well as where it will go, will keep you from running out of materials before you complete your project.

BUILDING YOUR WALL

Once you have planned where to build your wall, you can begin the building process. You can move a dry stone wall once it is in place, more so than a mortared wall, but it is always best to decide where you want it the first time. Walls under trees or too close to your house may hinder your ability to expand or add on to your home, and the trees may cause the stone wall to crumble when the roots push the foundation stones out of place.

MATERIALS YOU WILL NEED:

- Several tons of fieldstone depending on how long you want your wall.

> **NOTE:** The fieldstone is irregular, and an exact estimate cannot be done, but using the 6-inch per stone formula, if you choose to make a wall two courses thick, and you need a wall that is 200 feet long, you would need about 800 stones for one level of your wall. If you want your wall 3 feet high, you need to divide 36 by 6 (a 36-inch wall divided by 6 inches, which is the height of each stone) and then multiply by 800. This will be 4,800, which will be the number of stones you will need at the very least for this wall.

- Five hundred pounds or more of gravel and/or riprap for the hearting.

TOOLS YOU WILL NEED:

- Pick
- Shovel
- Hand spade
- Gloves
- Goggles
- Mason's hammer
- Chisel
- Sledgehammer

If it is a freestanding wall with no cornerstones for other adjoining walls, you will face your stones slightly inward to allow the wall to withstand natural movement. You will also want to use capstones to keep out debris, such as leaves, twigs, and seeds. Keep in mind you do not want to place a freestanding wall too close to trees because tree roots can eventually push the wall apart.

INSTRUCTIONS:

1. Using a shovel, dig out a portion of topsoil down on two sides to the more stable subsoil in an inverted V-shape with the point toward the middle of the trench, also called a footing ditch. This footing ditch should be at least 6 inches below the frost line for your area. **Note: You can find your frost line by checking with your local building inspector.**
2. Get rid of any debris or grass.
3. Begin by setting your stones in this trench, one on either side, letting them naturally slope toward each other.
4. Continue to do this all along the base of your wall.

If your wall is two stones wide, you can make it about 1 to 2 feet tall without any structural issues. For taller walls, such as ones 4 feet tall, your wall needs to be at least 2 feet wide. A standard dry stone wall is approximately 3 feet tall by 2 feet wide.

The dry stone retaining wall

The main purpose of a retaining wall is to keep an embankment from sliding because of erosion. Tall embankments require terracing with one or two terraced walls to make them functional, but smaller ones can be properly retained with just one wall two or three stones tall. Retaining walls should have an even thickness at the base and at the top, not like a freestanding dry stone wall, which uses smaller stones on the top than on the bottom.

MATERIALS YOU WILL NEED:

- Several tons of fieldstone — the amount will depend on the length and height of the wall.

A quick formula for estimating the amount of stones you will need is as follows:

1. Multiply the length — in inches or feet — of your wall by two.
2. Multiply this number again by two — if you are building a two-course wall.
3. Measure the height — in inches or feet — and divide by six.
4. Multiply this number by the number you arrived at for length.

- Five hundred pounds or more of gravel and/or riprap.
- Filter fabric — available at stores that carry landscaping or pool supplies.

TOOLS YOU WILL NEED:

- Pick
- Shovel
- Hand spade
- Gloves
- Mason's hammer
- Chisel
- Sledgehammer
- Goggles

Retaining walls are built into an embankment so you must dig away any loose soil at the base until you have firm soil. You will need anywhere from 18-inch to 24-inch stones at the base.

INSTRUCTIONS:

1. Begin by digging a trench at the base of the embankment where you want your wall. This trench should be about 2 feet wide and at least 6 to 8 inches deep. Fill

the trench with 1 to 2 inches of gravel. This trench is also called a footing ditch.

2. Start laying your heaviest stones at the base. You should lay the entire base of the wall before you begin to build the wall up.
3. As you begin to work the course up, cover all your vertical joints and slope your stones inward toward the embankment by setting them in at a slight angle toward the slope. You will want to dig the stones slightly — 1 to 2 inches — into the embankment as you build up. The soil will settle into the stones and provide strength.
4. Continue to place stones firmly on top of one another and into the earth's slope, digging each stone into the soil in as you go.
5. Place the heaviest capstones firmly on top of your wall.
6. Place your stones at least 1 or 2 inches above the embankment to accommodate the wash that will filter down. **Wash** is the accumulation of rainwater and organic matter, including soil, that will filter down through the rocks. If you do not place your stones higher than the embankment, the wash will flow out over the top stones and erode the wall of earth behind the retaining wall and eventually cause it to crumble.

It is advised to place a filter fabric along the top on the compacted earth. This is the same fabric sold in home hardware stores that keeps weeds from growing through flower beds. It allows water and loose soil to pass through without causing an erosive wash during heavy rains. You will not need to be replace it once it is in place.

Retaining walls taller than 3 feet will need curves or buttresses to support them properly. **Buttresses** are very large stones set into the wall to provide a strong center point along the wall. You need interlock them with the wall just like the other stones. When you build a retaining wall along a curve, buttresses are not as important because a curved wall is stronger than a long, straight one. The reason for this is simple: Curved walls allow for small movement of the earth and stone with gravity and natural forces, while a straight wall does not move with natural forces and needs a strong center stone.

Remember to backfill your wall with hearting because it will help keep the soil in place.

Terraced Walls

Terraced dry stone walls are multiple retaining walls used when the hill or embankment is steep and a single retaining wall made from stone would be weakened by gravity. These walls require a large amount of stone, and it is much more difficult to create terraced courses, primarily because you have to get stone up to each terrace to work with it. If at all possible, work from the top down when terracing.

MATERIALS YOU WILL NEED:

- Several tons of fieldstone — the amount will depend on the length and height of the wall.

A quick formula for estimation is as follows:

1. Multiply the length — in inches or feet — of your wall by two.
2. Multiply this number again by two — if you are building a two-course wall.
3. Measure the height — in inches or feet — and divide by six.
4. Multiply this number by the number you arrived at for length.

- Five hundred pounds or more of gravel and/or riprap.

TOOLS YOU WILL NEED:

- Pick
- Shovel
- Sledgehammer
- Mason's hammer
- Chisels
- Planks of wood for moving rock
- Wheelbarrow
- Goggles
- Gloves

You need a level footing on each step on which to work, and you can use this footing for garden plantings after you establish the walls.

Preparing the terraces

You can choose to have the hill or steep slope that will be terraced completely excavated mechanically with terraced areas plowed in. But, this will leave your slope bare and is very invasive to natural plantings, such as small pines or cedars, because excavating destroys these plantings, the seedlings, and the balance of the soil on which small cedars and pines grow. For safety's sake, it is also wiser to excavate — remove earth and organic materials — by hand using a pick and shovel to clear the soil to create each terraced area, allowing you to keep the natural plantings in place. You need multiple retaining walls because the slope is steep, and using mechanical equipment, such as tractors, can be very dangerous.

Laying the stone

The base depth of your foundation for each terrace should be at least half of your finished wall height. For example, if you want your terraces to climb up 4 feet of hillside, you will need to make each terrace's foundation at least 2 feet wide.

DIRECTIONS:

1. Begin by digging a trench, also called a footing ditch, at the base of the embankment where you want your wall. This trench should be about 2 feet wide and at least 6 to 8 inches deep.
2. Fill the trench with 1 to 2 inches of gravel.
3. Start laying your heaviest stones at the base. You should lay the entire base of the wall before you begin to build the wall up.
4. As you begin to work the course up, cover all your vertical joints and slope your stones inward toward the embankment. You will want to dig the stones slightly — 1 to 2 inches — into the embankment as you build up, using the pick and shovel to insert the stones into a wall as you create each terraced area. The earth will settle into the stones and provide strength.
5. Continue to place your stones firmly on top of one another and into the earth's slope, digging them in as you go.

6. After you have completed your first terrace, begin your second terrace in the same fashion.

7. Place your stones at least 1 or 2 inches above the topmost terrace to accommodate the wash that will filter down.

8. Bury all of your irregular stone sides into the wall, and keep the flatter surfaces toward the top for easy stacking. Again, always use the longest stones you can find and always cover your vertical joints. Just think strength because if nothing else, you must have strong retaining walls. Cutting into the hillside rather than providing an outward terrace will weaken the effectiveness of multiple retaining walls.

One of the primary benefits to having multiple terraced retaining walls is simple: If water has been creating erosion, the water will be slowed considerably by having multiple walls of rock to filter through, whereas one wall will not allow the slow draining of high runoff.

Plantings along the terrace will add interest and help keep erosion down. You should always use shallow-rooted shrubs, such as small cedars and Indian Hawthorne. Do not plant any type of oak trees because they have deep root structure that may create problems with the terraced walls in later years.

Building Veneered Walls

An outside entertainment area built using flagstone on the floor and stone walls and columns.

Using mortared wall techniques will help you accomplish building a wall with concrete block and then creating the look of solid stone on the surface of the concrete block. It can take considerably less time and effort than using actual stone and can also be more cost effective, but you will need a concrete footing, mortar, and stone that has at least one flat surface. The concrete block interior wall must be even and level, or your veneered stone will reflect the skew of the wall and you cannot correct it. It is important to note that most homes built from "solid" stone after the 1900s are actually veneer, or stone over concrete block or brick, which is the actual definition of the word. Building a wall of concrete and applying a veneer of manufactured or natural stone is just as aesthetically pleasing and solid as one made from actual stone.

Estimating the amount of block and mortar you will need

The industry standard for concrete blocks is 8 inches in height — in actuality, 7 5/8 inches to allow for the mortared joint — and approximately 16 inches long — in actuality, 15 5/8 inches — with either two or three large holes in the center.

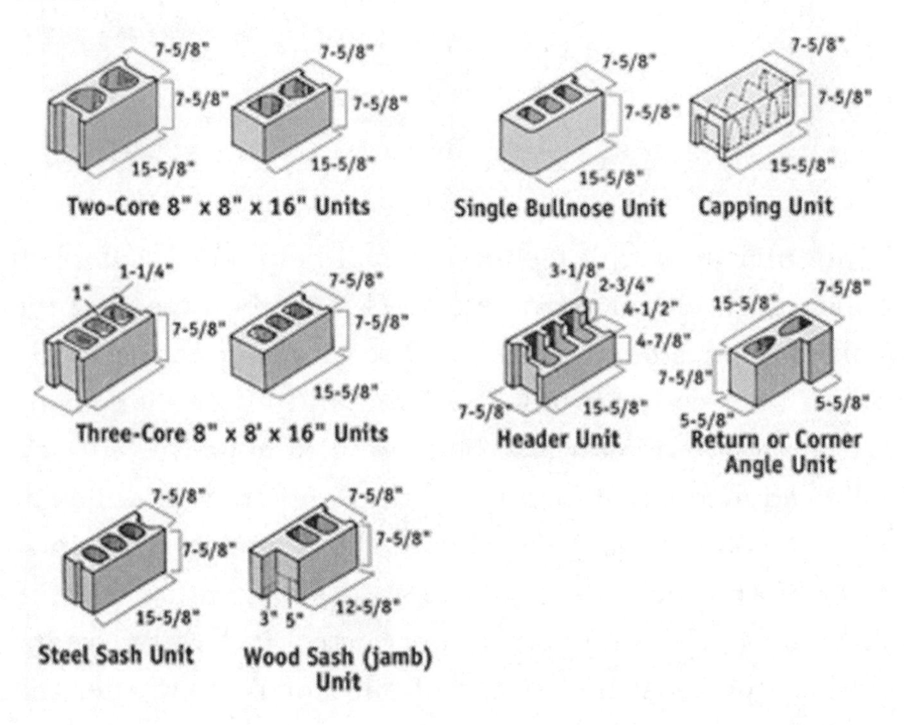

You should place every block wall on a poured concrete footing ditch or foundation trench, which should be twice as deep as the thickness of the wall and at least one block, or 8 inches, deep. To determine the amount of block needed, use these measurements by simply measuring out the length of wall you intend to build and how high you want it.

For instance, if you are building a wall 6 feet long (72 inches) by 4 feet high (48 inches), follow this equation:

1. Start by dividing 72 inches by 18 — the length of the concrete block. This will give you four; therefore, you need four 18-inch blocks for the base of the wall.
2. For that same wall to be 4 feet high, you will need to divide 48 inches by 8 inches — each block is 8 inches high — which results in six blocks.
3. Next, multiply the results from these equations, which is four blocks (the height) by six blocks (the length) to get a result of 24. This would be the minimum amount of block needed to complete the wall.

Pouring the footing for the concrete block wall

No matter what size block wall you build, it will require a secure foundation of poured concrete, much like the footing you needed to pour for the barbecue pit. You should always pour the footing for the block wall below the frost line. The foundation should be as deep as the thickness of the wall, and twice as wide. For example, if you are building a concrete wall one block thick (8 inches), then your footing should be 8 inches deep and 16 inches wide. This will give a 4-inch allowance on either side of the block wall when you are done building it. Digging this type of trench is called a footing ditch.

MAKING A FORM

In order to create a footing in a specific area, you will need to make a wooden form to keep the poured concrete contained within boundaries.

MATERIALS YOU WILL NEED:

- Approximately 15 two-by-four wooden boards.
- Fifty small wooden stakes.
- Thirty pounds gravel.

TOOLS YOU WILL NEED:

- Shovel
- Claw hammer

You can make this form using two-by-four wooden boards held in place with wooden stakes driven into the ground every 4 inches outside of the form. If you suspect a drainage problem in the area where you will build your wall, lay a drain line along the outside of the form. To build a drain line, simply dig a 1-inch deep trench line deeper than the foundation line. Backfill this trench with gravel or crushed stone. This will reroute any water buildup along the wall. If you are building a very long wall, make the forms into sections that you can handle, and pour small sections at a time. Complete each section separately.

Here is a cement slab with a cement block footer. The footer will be used to build cement block walls. Photo courtesy of Douglas R. Brown.

LEVELING THE FOUNDATION

Before you pour the foundation, you will need to make sure the footing remains level. To do this, place a level across your foundation approximately every 8 inches. Adjust the height of your two-by-fours by raising or lowering the stakes to assure the boards are level.

POURING THE FOUNDATION

1. Mix your concrete, and fill the form with the mixture up to the top edge of the wooden form.

2. Level the poured concrete by dragging a two-by-four across the top of the freshly poured concrete. Work the two-by-four back and forth in a zigzag fashion until the concrete is level.

3. Fill in any unlevel spots with more concrete, and wipe away any high spots with a two-by-four wooden board. Allow the concrete to cure and dry.

Concrete Block Wall with Stone Veneer

Again, building a concrete block wall as a base and adding stone veneer is a wonderful way to add beauty and function to your landscaping or your home's exterior or interior. As in any project, you will have to make adjustments depending on the width and length of the wall you intend to build.

MATERIALS YOU WILL NEED FOR A WALL THAT IS 4 FEET TALL BY 5 FEET LONG:

- Approximately 50 to 100 concrete blocks — use your own estimate depending on the size and height of the wall you are building.
- One to 2 tons of manufactured stone or brick, as rectangular as possible. If you are using natural stone, you will need to use thin fieldstone or flagstone. Large, heavy stones will not hold to the wall.
- Three 50 pound bags of Portland cement — or more depending on your project.
- One hundred pounds sand.
- One bag of hydrated lime — 50 pounds.
- One gallon or more of tile mastic, adhesive sold at hardware stores for setting tile. Purchase this only if

you have purchased thin manufactured veneer stone or brick or are using flagstone.

- Approximately ten to 15 wooden two-by-fours to make the form for the foundation.
- Approximately 30 to 50 small wooden stakes to keep the form in place.
- Masonry ties and (optional) anchor bolts if using wood on the top of the wall.

TOOLS YOU WILL NEED:

- Trowels (small and large)
- Shovel
- Wheelbarrow (for mixing cement and grout)
- Gloves
- Garden Hose
- Level
- Jointer (optional)
- Masonry hammer and chisel
- Mortarboard

INSTRUCTIONS:

1. Drive stakes into the ground, and string a straight line between the stakes to help keep your wall straight.

2. Dig a trench at least 6 inches deep for your foundation, which should be a poured concrete foundation.

3. Determine the amount of block you will need for the first course by laying the block down onto the foundation, leaving a 3/8-inch gap between each block to allow for the mortar. This will give you a good idea as to how many blocks you will actually need and could differ

from your estimate but normally not by many more than a few blocks.

4. Begin laying block by spreading mortar mix with your trowel about 1 inch deep and 8 inches wide on the concrete footing. Extend the mortar mix about three blocks. Using your trowel, furrow the mortar by creating a slight depression in the center of the laid mortar. This will force the mortar from the center of the block after it is laid to the edges. If you are building a wall with a corner, or two walls, lay your block from the corner first.

NOTE: Never mix more mortar than you can use in a two-hour span of time. You should mix mortar in a wheelbarrow and cover it with plastic to prevent moisture from evaporating too quickly.

5. Lay your block, starting from one end of your wall, or the corner, continuing to set the block into the mortar. Press down to form a tight seal. As you finish the base, begin to work the course up, overlapping all vertical joints. You may need to split the block as you go.

6. Scoop the mortar with your trowel as you go, placing all the unused mortar on a mortarboard, which is a large piece of hand-held wood that you can throw the excess mortar on as you work. You can continue to reuse wet mortar as long as it does not begin to cure and dry. Keep turning this mortar to prevent it from setting up too quickly before you can reuse it.

7. Make sure as you lay the block next to each other, you leave approximately 3/8 inch to ½ inch of mortar on the sides to ensure a tight seal to each block. As you lay the base block of the walls, check your level as you go, making adjustments in the mortar to ensure a level surface.

8. If you need to cut a block to end the wall within certain confines, use a mason's hammer and block chisel to split the block. Always tap the blocks into alignment, and do not attempt to move the blocks after the mortar has set.

9. Keep leveling and realigning the wall as you lay the block. Remember to continue to stagger all the blocks to cover all vertical joints. Always build the corners up first if you are creating a wall that has corners.

10. If you want a load-bearing wall, or in other words, if you will use the wall to hold planters or if people may walk along the wall, you may need to reinforce the block by inserting rebar into the holes of the block and using cement to hold the rebar and blocks in place. When you add veneer to a new concrete block wall, insert a small piece of reinforcing metal called masonry ties into the wet mortar every other block. These pieces of metal will help the veneer rock hold to the concrete and will not be seen once the veneer is in place. For an existing concrete wall, you can screw the masonry ties into the hardened mortar.

11. Finish the wall by removing all excess mortar and making the surface smooth for the placement of the stone veneer. You may want to use a tool called a jointer, which is a trowel similar to a pointing tool, to press the mortared joints in and tight. A jointer has squared or V-shaped edges that will affect how the mortar looks from the outside. Because you will add stone veneer to the outside, you will not see the concrete block.
12. If you will add a wooden top rather using capstones, insert anchor bolts in the top after you have filled the top course with concrete. This will allow you to securely bolt the wood to the top of the last course.

Adding the stone veneer

A 6-inch veneer of stone is the maximum depth you will want on the outside of your concrete wall because gravity will affect rock placed on the outside of concrete block, and it is harder to work with anything heavier than 6-inch thick veneer. Commercial stones made from concrete mimic the look of natural stone and can be very convincing if you take the time to mortar them in correctly. If you did not mortar in your masonry ties, you can anchor them to the exposed cement block by screwing them in with concrete screws and a masonry bit. You will bend these ties around the stones as you place them on the wall, and you will not see them from the outside.

INSTRUCTIONS:

1. Lay the veneer stone as you would any other stone — only vertically attaching it to the concrete wall. You will not stack the stone horizontally but vertically. Use a bed of mortar about ½ inch thick against the wall, and using the methods of dry stone stacking, lay each stone from the ground up.

2. Pack mortar in between the stones, and use your pointing tool to press the mortar in tightly. This technique is used because although you want to mimic the look of a dry stone stack, you need to make sure the veneer is strongly attached to the block, or it may pop off in extreme weather conditions. You can recess the joints deep into the stone crevices, but you still need some sort of mortar joint for strength.

3. When doing the veneer on the block, go up only three blocks high before letting the wall cure for three days, keeping it wet a couple of times each day by either misting with a garden hose or covering with a tarp and letting the natural moisture build with heat. The water allows the mortar to cure properly.

4. If you have purchased very thin manufactured veneer stone, you can use commercial tile mastic to hold the stone to the concrete wall rather than mortar, and use grout to create joint lines, much like you would if you were using tile.

Now that you have advanced to building walls, patios, and barbecues in your landscaping, you may be ready to take on the challenges of building a rock or water garden. These projects will make use of all the knowledge you have gained in creating your other projects and will include using preformed pools and liners to create entire settings with plants and pathways.

CHAPTER 7:

Creating Rock Gardens

Rock gardens come in two varieties. The first variety is the **dry rock garden**, which can contain sun-loving plants, such as succulents, impatiens, and moss roses for color. Dry rock gardens are popular with many landscapers because they require little or no upkeep and provide a pleasant garden feature for your enjoyment. Another form of rock gardens is the **water garden**, which uses ponds and water as part of the natural elements. These gardens normally include a rock waterfall. Rock and water gardens have accented landscapes for hundreds of years. The Japanese had become masters of creating rock and water gardens back in A.D. 785, long before Colonial America had this landscaping. Gardens using rock alone became enormously popular in Victorian-era America, primarily because of their easy maintenance.

Adding a water element to your landscape, or creating a rock and water garden, can be done without disturbing too much of the original rock garden. A water garden provides the avid gardener with the opportunity to add plants not normally seen in traditional gardens. Depending on the region of the country you live in, you may not have access to succulents like aloe and cactus that you can easily introduce to your landscape in rock gardens. If you live in an arid region, you may want to use rock jasmine or alpines, but be prepared to devote time for additional watering, and plant these in a semi-shaded area.

Some botanical gardens, such as the Chelsea Physic Gardens in Chelsea, England, collect native rock from areas such as Iceland to help the propagation of the plants that grew from that region. **Rockeries**, which are gardens primarily made from small rock and rubble, complement the plants, and the use of native stone added a more historically correct adaptation of the natural environment in the Chelsea gardens. The rock garden is now seeing a resurrection in popularity as a garden accent because it is easy to maintain, adds an aesthetic element, and creates a peaceful and interesting area in landscaping.

Building a Dry Rock Garden

Small rock garden.

If your space or resources are limited, you can still install a small rock garden in a variety of different looks. There are six basic types of dry rock gardens:

- Sloping outcrop — resembles a cliffside.
- Sloping terrace — same as a sloping outcrop but with larger, flatter top surfaces.
- Pavement — flagstone stepping stones with earth in between to support plant life.
- Cliffside (or bluff) — resembles a stone wall with an earth top surface.
- Gorge — stone wall built on two sides with area for plants between.

- Cropping — a mound of rocks piled into one area and surrounded with trees and shrubs with flowering rock garden plantings within the cropping.

You can easily make dry rock gardens by creating a large stone cropping, confining the grouping to a certain area, and using plantings to enhance the stone and landscape. Creating pockets between stones for plants to grow adds interest to your rock garden.

Plan the area

Planning your rock garden is the first step in building it. You should aim to use rocks and stone that produce a natural-looking setting within your landscape. Draw a rough sketch of how you see your rock garden fitting into your yard. Once you have placed all your rocks, you do not want to have to move them again, especially if you have completed planting your flowers and shrubbery. You should also plan for a possible water and rock garden in the same location in case you decide to add water as a feature in your garden. Try not to place your garden under trees that lose leaves during the fall and winter months or where excess root growth occurs because more leaves on the ground and in the rocks create more maintenance, and excessive tree roots may cause movement and damage to your existing rock garden over the years. If you are going to use water as a feature later, or if you plan on using shade-loving plants, you will also need to be careful not to place your rock garden in full sun.

WORK WITH A CLEAN SPACE

Clear all debris, trees, bushes, undergrowth, and shrubs from the area. Cut any low, overhanging branches to allow you to work in the area without hurting yourself. Your goal is to separate the lawn section of your landscape from the rock garden area, although in time it will grow to blend in naturally. Create a deep edge along the rim of where you would like the stone garden, about 3 inches deep, to help define this initial separation. If you plan to use curved sections, you may want to mark the area with paint to help you visualize how the garden will flow. This also helps keep you on track as to rock placement.

INSTALL THE LARGEST STONES FIRST

Rock gardens should have a mix of large and small stones. You may also consider a pathway intertwining your rock garden. Set your largest stones where you want them, using crowbars and shims to roll them in place. If you want a small pathway or a dry river bed made from gravel, plan where this will interact with the large boulders. Either mark this area with the paint or hoe it out to accommodate gravel and flagstone.

SET THE SMALLER STONES

Set the next size of stones down from the large boulders, and start placing them among the large stones. Create a base of the largest stones first, and then begin to stack them, letting them naturally fall in to place. Use gravel and riprap to fill in small pockets, and you can use peat moss or rich loam, which is a form of soil composed of sand, clay, and silt, for plantings in the larger pockets. Install gravel along the dry river bed, and create your pathway. Remember, rock gardens should look like natural outcroppings of stone. This is not a formal garden so any design goes.

SET THE PLANTS

For a better survival rate, choose plants suited to your natural climate. **Xeriscaping** in rock gardens, which is using plants that require little water other than the rainfall in your region, is a wise method since you may not have placed your rock garden near a water source. Rocks also tend to draw moisture from plants and from the surrounding soil, and your plants may dry out more rapidly. You can use annuals and perennials — annuals being plants that continue growing throughout the year, and perennials being plants that are replaced twice a year as the seasons change to add color and interest. You may also want to use mulch to help the plants retain water and to help maintain a neat appearance.

Rock garden plants

This is certainly not an all-inclusive list, but these plants are readily available in all regions and can be used in dry rock gardens with optimum success. These plants are hardy and require little care.

New Zealand Burr (acaena)	Carpeting perennial
Japanese Maple (acer)	Compact, slow-growing tree
Alpine Yellow (achillea)	Crevice and crack filler
Aethionema (aethionema)	Low-growing shrubbery; evergreen
Bugle (ajuga)	Ground cover; crevice and crack filler
Flowering Garlic (allium)	Tall, background plant
Alyssum (alyssum)	Flowering perennial
Mount Atlas Daisy (anacyclus)	Crevice and crack filler; flowering
Rock Jasmine (androsace)	Flowering perennial
Artemesia (artemesia)	Gray-green filler
Snow-In-Summer (cerastium)	Flowering perennial
Twinspur (diascia)	Sprawling, flowering perennial

Building a Water Garden

Again, planning is essential when deciding where to place your water garden. You may also want to consider whether you want to include fish in your water garden. You certainly do not want trees with leaves hanging over the water; the leaves will clog your filter and cause more maintenance than will already be necessary. Most

This water garden has plenty of plants for fish to hide under. Photo courtesy of Douglas R. Brown.

water gardens have a waterfall, which also acts as a filter and adds oxygen for plant and animal life. Water gardens are thriving environments that make noise and attract wildlife and insects. Raccoons will discover your water garden rather quickly and may even eat fish from the pond so you need water lilies and under growing plants for fish to hide in. Primarily predatory insects, such as dragonflies and water beetles, will come to your water garden, and while they do not cause problems with ponds, mosquitoes and their larvae do. Even though dragonflies and beetles will eat much of the pesky insects, you may want to put in a commercially made tablet that prohibits the growth of larvae but will not harm fish or other wildlife. Another option is larvicides, such as mosquito dunks, which are donut-shaped bacteria tablets you drop in fish ponds to kill mosquito larvae. They contain no poisonous chemicals and are harmless to birds and pond fish, and one tablet treats 100 square feet of standing water for approximately one month.

Both examples of a pond utilizing natural stone with a waterfall and koi fish.

As you can see, water gardens do require a fair amount of upkeep. You must keep the pond clear of leaves, keep the fish fed daily and warm during the winter months, and keep the plants rooted and trimmed as they grow. You always need to clean filters, and sometimes you need to replace the water if algae growth becomes rampant. The benefits of having a beautifully maintained water garden are numerous. Water gardens make lovely sounds as the water pours from the rocks, and the sound is reported to reduce stress and anxiety; therefore, you may want to place it near your screen room or patio. Water gardens are beautiful to look at and watching insects and wildlife interact can be fascinating for both young and old visitors to your garden. Birds love to congregate near water gardens, to bathe and eat insects, and you may want to consider incorporating a bird feeder or two near your water garden.

Where to put your water garden

There are several things you will need to consider when deciding where to place a water garden. You should consider the following points before you begin working on this project:

- Aquatic plants need at least four hours of direct sun every day to thrive so you should place your water garden where it receives at least that amount of light. If you want to put water lilies in your pond, they require a bit more sun, usually six hours or more.

- The area you choose should have good drainage. You also do not want pesticides or debris washing down into your pond so be careful if you are placing the water garden on a slope that may have plantings or excess top soil.

Choose your pond location by using the following points. *Please see pages 194 and 196 for the project construction.*

- If you plan to connect two ponds with a stream, you will want to have some slope for natural gravity to help keep the stream flowing, but this does not require a large amount of slope since the water is pushed into the stream area via the filter.
- Trees have leaves and roots, neither of which is good for your water garden. Be sure you consider where trees are located before placing your water garden.

- Check with your local building code officer through the city offices, as well as your licensing and permitting department, about whether you will need to build a fence or barrier around your water garden to prevent small animals or children from possibly drowning.
- You will need to install a pump, filter, and fountain to keep the water circulating and fresh. If you are incorporating fish into your water garden, you will need a biological filter. A biological filter adds beneficial bacteria to break down fish waste and other organic matter. Because these components are necessary, you will need access to electricity and water.

This bridge covers the return area for the water being pumped from the pond up the hill into the stream, which flows back into the pond as it forms a small waterfall.

A photo of the redwood bridge and seating/observation area that overlooks the stream and pond.

Establishing the water basin

There are a couple of ways to establish the water basin, which is the actual area where the water will be contained. The first way: use a flexible, heavy-duty, rubber-like liner used for installing flat roofs on commercial buildings. Dig the shape and size pond you want, and then line the inside with the rubber liner. By using this method, you create the ability to dig larger, deeper ponds capable of holding many more plants and fish that the prefabricated plastic liners.

If you are interested in making a water garden of a smaller size, the other method is to use preformed, rigid shells of heavy-duty polyethylene or fiberglass, which are sold in most home hardware stores. Most of these preformed shells have built-in ledges for plants and do not require as much ground preparation as the flexible liners, but using preformed plastic shells limit your design to whatever may be available to you. Additionally, you can buy two or three of the shells and link them together with piped streams, creating an interesting effect in your garden. Most of the shells have molded spillways, which are actual downward-pointed plastic pipings that you can use to create streams from one pond to the other.

A rubber pond liner is being positioned for placement for the stream bed, waterfall and pond. *A before image can viewed on page 192.* Photo courtesy of Douglas R. Brown

Beginning the water garden

Determine the size of your pond using flexible tubing or a garden hose to form the shape. This will also determine the amount of water you will need to fill the pond. For instance, a pond dug 2 feet deep, which will only support a small amount of fish and plant life, and 11 feet in diameter will hold approximately 1,426 gallons of water. A pond 4 feet deep with an 8-foot diameter will hold 1,508 gallons of water. Larger ponds are easier to maintain and are more stable than smaller ones and will support the larger koi and goldfish. You may also have to dig below the frost line in colder climates, which means a greater depth. After you have laid out the shape and determined your depth, you can estimate what size pond liner you will need.

Estimating the liner

Water garden experts use the following formula to determine the size of pond liner needed for water gardens:

- Length — add overall length plus twice the maximum depth.
- Width — add the overall width plus twice the maximum depth.

It is a good idea to add a foot or two to both of these measurements to allow for error and to give the pond more overlap around the edges.

Example: A pond 6 feet wide, 11 feet long, and 1½ feet deep will need a liner 10 by 15.

Width: 6 + 1.5 + 1.5 + 1 = 10

Length: 11 + 1.5 + 1.5 + 1 = 15

When you dig your pond, you will need to add plant shelving that you will submerge in the water either fully or partially along the inside edges to support potted plants at varying levels. You will have plants that require no more than 18 inches of room to grow, some that require 24 inches, and fish that will require a deeper level of water at least 2 to 3 feet. Plant shelving can be earthen terraces underneath the rubber liner; block or stone built up to certain levels; or simply terra cotta pots placed upside down to support other pots with plantings.

Gravel and rocks are being placed on the rubber pond liner for the stream bed, waterfall, and pond. *A before image can viewed on page 192.*

Placing the pond skimmer

Pond skimmers are filters that can range in price from $100 to $700 or more, depending on the size of your pond. These devices actually have a flat lip that draws the pond water in, filters the debris, and releases the cleaned water out. There are also weirs, which are small dams that allow water flow but prevent debris and animal life from entering or exiting. You can install weirs in the streams to keep fish from going into a lower pond. You should bury the skimmer to the proper level beside the pond — the manufacturer's instructions will help to determine that depth. Dig a ditch to the external pond pump and from the pump to the external pond filter. If you are using a submersible pump inside the pond skimmer, then dig the ditch from the skimmer to the external pond filter.

A pond skimmer and pump that is disguised by an artificial rock. Photo courtesy of Douglas R. Brown.

Placing the pond underlayment

Pond underlayment is a heavy, synthetic matting used to protect the liner from roots and rocks. You can purchase pond underlayment from most home hardware stores and pool suppliers. Even the smallest stone can eventually wear a hole in the liner, which would mean you have to replace it; therefore, it is best to take your time cleaning the area you have dug out for the pond to remove any unwanted rocks or debris. Once you have dug the pond, place the pond underlayment in the hole and overlap along the outside if you have enough material. Any place the liner touches should have underlayment. If you need to use more than one piece, or have smaller pieces you need to place together, tape them with duct tape to prevent it from moving when the pond liner is placed.

Rocks are being placed on top of the stream bed liner. Photo courtesy of Douglas R. Brown.

Placing the pond liner

Position the pond liner over the underlayment as evenly as possible. Some overlap and wrinkling can occur, but these should flatten out once you add water. Remember to use at least 1 foot of overlap around the pond's edges.

A solar light designed to look like a rock along the stream bed.

The finished waterfall spilling into the pond.

Excavating for streams and waterfalls

An external pond filter or waterfall tank, which are both filters that work outside the pond itself and create a waterfall effect, can be placed above the pond to allow water to flow from one pond to another or from a stack of rocks into the pond. Position this filter to spill directly into the pond. You will stack stone along the edges and the face of the filter to create the waterfall. If you want a stream and you want the stream to spill into another pond, you will need to dig the stream bed and the second pond to their desired depths, sizes, and shapes. If you position your initial pond at the top of a slope, you can add plumbing and a second pond later. Remember to dig the stream wider than you want the finished product because you will fill the bed with stone and gravel.

You can secure stones to the filter and to each other by using expandable foam available through home hardware stores or by using mortar. Stack the stones to disguise the filter, but allow the water to flow from the top down along the stones and into the pond. You will have to adjust the stone accordingly to achieve this effect. Connect all of the plumbing per the manufacturer's instructions.

Large stones were used to create a wall for a backyard pond.

In this photo, you can see that different types and sizes of rocks were used to create a border. Photo courtesy of Douglas R. Brown.

Build your rock garden

Placing stones around the edges of your pond is easy enough to do, but avoid creating a necklace effect, which is unflattering and unnatural. The necklace effect occurs when you place stones around your pond in an even fashion and when you use stones approximately the same size. Natural ponds have various sized rocks that surround them, and creek beds always have small and large rocks both outside the stream bed and within. To avoid this necklace effect, make large groupings of various sizes and shapes of stone, and intersperse with smaller groupings to give the effect of stone outcroppings. Add stone as close to the edge of the pond as you can along the flat overlaps of the liner. You can also use large, flat stones, such as flagstones, to overlap the actual edge of the pond and stack other stones on top of them. If people will walk along the edge, you will need to mortar these stones in place. If not, you do not need to use mortar. Fill your pond with water from the garden hose, and add dechlorinator, which is found at any aquarium or pool supply store and is used to remove any chlorine and chloramine and to help the pond establish the proper biological balance to allow plant and animal life to thrive.

Use both small and large rocks inside the stream to form natural dams.

Large and small rocks work well on the outside of the pond to cover the rubber liner.

Adding your water plants

Add aquatic plants as soon as possible after you complete the last phases of building your stone water garden and after you add water and dechlorinate it. Many popular aquatic plants, such as dwarf cattails and arrowroot, actually use the nutrients that would otherwise feed algae and pond scum, thereby helping to keep the pond clear. You can also add water lilies, which have large surface leaves that create shade for fish in full sun. Do not add more than 50 percent of surface leaf plants to your pond, especially if you do not have full sun for more than eight hours. The general rule is surface leaves should not cover more than one-third of the surface of the pond. This allows the sun to filter down to the plants below and helps maintain the balance of nutrients.

Lilly pads create shade for fish in full sun.

Water garden plants surround this waterfall.

You may also want to add packaged bacteria that are sold at water gardening stores to help start and maintain a clean and healthy pond. Add pond fish two at a time when establishing your pond to allow these fish to help biologically balance the pond to its optimal levels.

There are six basic types of water garden pond plants that are unusual compared to land plants because the plants do not breathe the same way. Typically, these specific types of plants breathe in oxygen during the evening and emit carbon dioxide, which is the reverse of most other chlorophyll-producing plants. Plants that live inside your home, for example, give off oxygen and take in carbon dioxide, cleaning the air. These special plants also convert carbon dioxide from the atmosphere and release oxygen into the water, which is why you will see small bubbles of oxygen on the submerged leaves on sunny days. They help provide oxygen for the fish; discourage the growth of algae, which do not thrive in oxygen-rich environments; and provide oxygen for other plant life.

The six basic types of water garden plants each serve a different purpose in the water garden.

- Water lilies — these deep-water plants will root strongly in the bottom of the pond, sometimes as deep as 6 feet, and have surface flowers and leaves that can be invasive if left uncontrolled. If you plant them in pots placed at the deepest end of the pond, they will be much easier to control.

- Oxygenators — these help maintain the biological balance of your pond by adding oxygen to the water, which discourages the growth of algae and pond scum. Fish spawn in these submerged plants, and insects and snails deposit their eggs and larvae on these plants, feeding your fish.

- Floating — these plants move freely in the pond and float along the water's surface.

- Partly emerging — these secure the underwater soil using their roots and prevent other aquatic life from rooting out the mud. They will add interest to your pond because they flower and attract butterflies and bees.

- Marginals — many of these plants have showy flowers and are similar to bog plants (*described below*).
- Bogs — while these cannot remain in constant water, they like moist soil. Their flowers and leaves are always above the surface, and shallow marginals are also called bog plants.

The rule for the number of plants you should place in your pond: For each 10 square feet, you should have two bunches of oxygenators, one bog plant, and one water lily. Bear in mind that aquatic plants can grow large quite quickly so you do not want to plant too many plants.

Water garden plant varieties

This is not an all-inclusive list, but these are readily available plants you can use in planting water or pond gardens:

Dwarf Cattail (Typha minima)	(Marginal) Perfect for small water gardens; survives low temperatures
King Tut Papyrus (Cyperus papyrus)	Also known as umbrella plant; grows in full shade or partial sun
Sweet Flag Water Grass (Acorus calamus)	(Marginal: bog) Iris-like foliage with yellow-green flowers in the summer; has a sweet smell when leaves are crushed
Juncus "Big Twister" (Juncus)	(Bog) Has unique twisted spiral foliage; prefers hot, sunny conditions

Marsh Marigold (Caltha palustris)	(Marginal: Partly emerging) Beautiful bright yellow flowers resembling marigolds; can be planted in water up to 4 inches in depth; requires full sun
Water Lilies (Nymphaea)	(Floating: Water Lilies) Water lilies, despite their fragile appearance, are incredibly hardy, come in many sizes and colors, and provide surface shade in the pond; most varieties are hardy to minus 30 degrees Fahrenheit

CHAPTER 8: Using Stone to Build Bridges and Dams

Building bridges to get people, animals, and vehicles across a span of water is one of the oldest ways in which people used stone. Using stones to build bridges started with a single stone set in a small stream to step on to avoid getting wet and evolved to the massive stonework involved in bridges, such as the original London Bridge that Peter de Colechurch built, which was the first bridge to span the Thames River. A 150-year-old stone bridge built at the Manassas Civil War Battlefield in Virginia carried soldiers and artillery cannons across the river, and it employed massive arches and buttresses for strength. These provide just a few examples of the long, historic use of stone to form bridges and dams.

A Brief History of Bridge Building

The early Romans built some of the most impressive stone-constructed bridges, and many have still survived, such as the Pons Fabricius, now called the Ponte dei Quattro Capi, in Rome. The Romans built this arched bridge in 62 B.C., and it has two semicircular arches that the Romans are credited with inventing, each spanning a massive 78 feet. They built this bridge with an upper arch built between the two arches to allow the release of water during the possible event of a flood, thereby maintaining the integrity of the bridge itself. Arches are also called **compression structures** because they rely on compression of two sides on a keystone, which is a large center stone put in place to keep the two arched sides from collapsing.

The Romans' efficient bridge building meant that future generations did not have to build any bridges for hundreds of years. When people started to move more within their countries and began developing societies, it became necessary to build more bridges. Sometime during the 12th century, Catholic priests and hired professionals commissioned the building of bridges. In France, a new order of priests who called themselves the Fréres du Pont designed and built bridges, the most famous of which was the Pont d' Avignon. This bridge was built in 1177 and spanned the Rhône River. This bridge had 22 stone arches, the longest being 115 feet from one end to the other.

You do not have to build a bridge spanning rivers or lakes. This section will teach you how to build a simple stone bridge you can use over water or over any place on your landscape.

Building a basic stone bridge

A simple stone bridge is a slab of stone with smaller stones supporting it built over a culvert or stream. The support stones need to be embedded into the soil on both ends of the culvert, which is a natural ditch that fills with water when heavy rains occur. The support stones form a level base for the slab to sit upon.

MATERIALS YOU WILL NEED (THIS IS FOR A SMALL BRIDGE, 36 INCHES BY 36 INCHES):

- One large stone slab, at least 36 inches long and wide and at least 6 inches thick.
- Approximately 1 ton of small stones — 6 inches by 6 inches — preferably ones with flat surfaces.

TOOLS YOU WILL NEED:

- Pick
- Shovel
- Hand spade
- Gloves
- Goggles
- Mason's hammer
- Level

> **NOTE:** Lifting a large stone slab of this nature can be too much for one person. If possible, get a couple of people to help you set this slab, or make use of a wooden tripod with a ratchet hoist — *instructions on making one on following page.*

INSTRUCTIONS:

1. On either side of the culvert or stream, dig two level trenches parallel to each other using your pick and shovel.

2. Lay the stones into this trench, making sure the leveled stone will be no more than 3 inches above the ground. Use the level to make sure the stones are sitting balanced.

3. Lay the big stone slab across the culvert or stream, resting either end on each stone "step" you have built from the instruction in step one.

4. Build the soil up around the stone slab and up over the steps, adding gravel if necessary, to create a ridge of soil connecting the slab to the surface of the earth.

Making a Ratchet Hoist

If you cannot find help to lift the stone slab, you can set it in place by making a simple ratchet hoist, which allows you to lift a heavy object by moving it with pulleys, hooks, and straps using the principles of leverage.

MATERIALS YOU WILL NEED:

- Three pieces of 2 by 6 hardwood lumber.
- Anchor bolts.
- Large hook and S-hook.
- Moving straps.

TOOLS YOU WILL NEED:

- Electric drill and bit — same size as anchor bolts

INSTRUCTIONS:

1. Anchor the three pieces of lumber in a tripod fashion using the electric drill and anchor bolts to attach them together.

2. Set the tripod in place over the stream, making sure the legs are securely sitting on either side of the stream.

3. Screw in the large hook, which will hold the ratchet hoist.

4. Attach the ratchet hoist with a large S-hook.

5. Attach the moving straps to the ratchet hoist.

6. Place the moving straps securely around the large stone slab.

7. Hoist the slab in place over the culvert or stream.

Building a basic arched bridge

Arched bridges are more complex and require more materials compared to a basic stone bridge, but they are more visually appealing. Ideally, an arched bridge is built over a larger span of water. Three basic styles of arched openings for bridges include:

- Semicircular arch — the most common and strongest arched bridge opening. The Romans invented this style, and people have used it for centuries to bear the weight of bridges. In a semicircular stone arch, all of the weight from each stone transfers to all sides respectively, and no one area or stone holds the most weight.

- Triangular arch — this bridge contains two side stones sloped together so the weight bears against, and onto, the side stones.
- Gothic arch — a primarily ornamental arch, yet still strong. Gothic arches have narrow, vertical, high sides that come to a curved point.

You can shape true arches into semicircles, ellipses, and catenaries (very pronounced arches not used for bridges but for interior and exterior design and sculpture, such as the St. Louis Arch). Rim-type arches are very slight arches commonly found on fireplaces and window openings.

Building a mortared, semicircular arched stone bridge approximately 36 inches wide and 5 feet long will require the materials, tools, and instructions listed below. This project should take no more than a week to complete.

MATERIALS YOU WILL NEED (THIS WILL MAKE A 5 FOOT LONG, SEMI-CIRCULAR ARCHED BRIDGE 36 INCHES TALL X 36 INCHES WIDE):

- One ton of squared stones, preferably sandstone or limestone, each approximately 12 inches wide, 8 inches thick, and up to 18 inches long.

- Two 20-foot-long pieces of ½-inch rebar and safety wire. If possible, have the rebar precut to 36-inch pieces.
- One 4-foot-by-8-foot sheet of ¾-inch construction-grade plywood.
- Thirty feet of 1-by-4 #2 pine boards, each cut 5 feet long.
- Four wooden stakes, each 24 inches long to use in forms and braces.

FOR THE MORTAR:

- Six bags of Portland cement.
- Two bags of lime.
- Two tons of sand.
- Access to water (garden hose).

FOR THE CONCRETE FOOTING:

- Six bags of Portland cement.
- One ton of gravel.
- One ton of sand.
- Access to water (garden hose).

TOOLS YOU WILL NEED:

- Pick
- Shovel
- Wheelbarrow
- Gloves
- Cement or mortar pan
- Hoe
- Mason's hammer
- Claw hammer
- Circular saw
- Trowel
- Pointing tool
- Stone chisel

INSTRUCTIONS:

1. Bridges are built on soft soil surrounding streams and culverts and need a footing. The larger and more complex the bridge, the bigger and more stable a foundation needs to be. Dig two trenches, also called footing ditches, 24 inches wide and 48 inches long on either side of the water source, such as a river, stream, pool, etc., to below the frost line, approximately 21 inches deep. You should make the trenches parallel to one another. **Note: You can find your frost line by checking with your local building inspector.**

2. Mix enough cement to fill both ditches to within 18 inches of the top. The basic recipe combines two shovels of cement, four of sand, and six of gravel in your wheelbarrow or mortar pan, adding water as necessary to make the cement a thick concrete that has the consistency of pancake batter.

3. Cut 12 lengths of rebar, unless they are already cut, 36 inches long, and set two into the cement of each ditch 8 inches apart from each other. Lay additional rebar — at least three pieces cut to 24 inches long — across the longer pieces. You can wire these together using any construction-grade wire, overlapping one wire over the other encasing the rebar, but once you have poured the cement, the structure will be encased. Most stonemasons recommend wiring the rebar to provide additional support for the bridge.

4. Place three pieces of upright rebar into the cement in the middle of the trench, and pound them into the earth, leaving each piece of rebar to project approximately 6 to 8 inches above the cement.

5. Pour an additional 3 inches of cement on top of the existing footing. Let this cement cure and dry for about an hour.

6. Set a 2-by-15-inch board in an upright position approximately 12 inches away from the back wall of each footing ditch. Brace the board tightly by using the stakes pounded into the soil at each end of the boards. You need this form to create a back "step," which will include the rebar that projected from the top.

7. Splash a little water on the hardened cement, and pour the remaining cement over the rebar but not over the board form. This will create an L-shaped section. Keep these sections wet, let them cure for two days, and then remove the boards. Let the cement cure and harden an additional two days.

8. Lay out and cut two arch forms from the plywood. You should cut these into a curve 12 inches tall and 48 inches wide.

9. Set these arches on either edge of the concrete footing, and nail a 1-by-4 board every 4 inches across both arches, connecting the arches. Use wooden shims

under the arch to wedge it in place on the concrete footing.

10. Mix your mortar, and begin laying the stone up against the vertical wall of the L-shaped foundation. You should use stones as flat on all ends as possible, and you should use as much mortar as necessary on the underside, but be aware that the outside of the stone will be visible so try to keep those mortar joints tidy. When you lay the second row of stones, overlap the existing joints.

11. When you reach the center, you will need wider keystones, also called arch stones. These will provide the strength that will keep the arch in place.

12. You can build up around the arch in a horizontal fashion to provide a rail for the walkway. You can do this before or after you remove the wooden forms from the arch. Either way, you must keep the structure damp for at least two days to allow the mortar to cure, and then allow it to dry.

13. Build up the soil around either side of the waterway around the stones to ease the transition from the flat earth to the stones.

Building a Dam

If you are damming a stream or creek that flows onto public lands or other property, you may need to obtain a dam permit. Small dams that create a minor impact are not usually subject to regulation nor is any stream or creek that is totally man made and on your own property. However, anything that affects landowners downstream from your residence needs a permit.

Here are instructions on how to obtain a permit:

1. You need to get a copy of the U.S. Geological Survey (USGS) quadrangle map for your land or creek site — providing the dam on your creek or stream will affect other lands, especially wetlands, as in the Florida Everglades or Carolina marshes, where damming creeks and rivers is strictly prohibited at any level. You can get these maps off the USGS website, **www.usgs.gov**, for a minimal fee.

2. Check the maps for any wetland or other sensitive ecological areas on the property you are planning the project. You can research these areas by accessing the U.S. Environmental Protection Agency (EPA) at **www.epa.gov**. Go to your individual state profiles for specific laws regulating the damming of waterways. If your dam affects waterways on federal land, you will also need to contact the district office of the Army Corps of Engineers to determine if you will need a federal permit, especially if your dam affects navigable waterways.

3. You will need to make sure you have accurate and up-to-date plans of your proposed project and clear

definitions of how and when you will complete the dam. Depending on the scope of your project and the affect on other areas, an inspector with the state agency may need to visit the proposed dam site. Make sure you keep all paperwork together with reference to the project and get everything in writing.

Building a simple stone dam

Why should you build a dam? Well, by definition, dams are blockages created to retain water in a specified location. Floodgates and levees are built to manage water flow. Natural dams are created when sticks and debris from a stream or river wash down the river, creating a blockage but still allowing water to seep through. Beavers build dams to create lodges for housing and to retain water into small pools that allow them to trap fish and other underwater wildlife on which they feed.

Building a dam from stone in a flowing river or stream would create a pool of water in your landscape. If you have already created small water gardens or streams, you may want to build yet another water feature such as a dam that will terrace your water down into another pool. Or, perhaps you have a stream and want to create a living pool as an accent in your landscape — the perfect reason for building a dam — and both the stream and dam will be healthy and vibrant water features. Adding a natural pool by damming a creek or stream and diverting water to the pool will draw birds and other wildlife to your pool.

Building a dam uses all the same techniques as building a retaining wall. However, most dams do not require mortar, and for this wall, you will only use a dry stacking technique.

MATERIALS YOU WILL NEED:

- At least half a pickup truck full of natural stones, depending on the size of stream or creek you will work with, such as limestone, fieldstone, granite, or sandstone.
- Five or more 2-by-6 or 2-by-10 pieces of pressure-treated lumber. You will need this lumber for bracing the rocks in the stream so if it is a large stream, you may need more than five pieces.
- Four two-by-four pieces of pressure-treated lumber.

TOOLS YOU WILL NEED:

- Shovel
- Hand spade
- Mason's hammer
- Chisel
- Waders or waterproof boots
- Circular saw
- Sledgehammer

INSTRUCTIONS:

You will need to divert the water from the stream off to another area of your landscape to build the dam. You can do this by creating a small stream by taking your shovel to the bank of the stream and digging a ditch for the spillover away from the existing streambed. If your small stream flows only during heavy rains or when snow melts, you may possibly build the dam without having to divert the water so plan on building your dam during the dry season.

1. Measure the width of the creek or stream. This will help you determine how many pieces of pressure-treated lumber you will need to effectively delay or stop the water flow. For example, for a 5-foot wide riverbed, you will need to have your 2-by-10 boards cut into 5-foot pieces.

2. Determine the height of your dam. For a 3-foot tall dam that uses 2-by-10 boards, you will need to divide 10 inches into 36 inches, which means you will need 3.6, or four, boards. Keep in mind that if you want the water to flow into a pond or pool, you may want to build the dam slightly lower than the river at its highest point so water will flow into the pond.

3. Using a sledgehammer, drive two two-by-four boards into each side of the creek, at least 1 foot deep and at least 4 to 6 inches apart into the creek bed. You should drive these into the creek bed in the opposite direction of the flow of the water. These will act as a brace for the stones that you will place in the dam.

4. Place your lumber boards in between the braces. Stack your stones as tight as you can on either side of the lumber wall.

You may be able to use scrap lumber from construction sites as the lumber you need will only act as an inner brace for the rock. The water will begin to settle the stones, and earth and debris will build up on the flow side of the stream, creating impaction for the stones. Impaction means the stones will become more fixed as the soil begins to act as an adherent.

Once you have constructed your dam, you may want to work on additional stone projects for your home.

Rocks of different sizes are used to create small areas of back flow and spillovers. Photo courtesy of Douglas R. Brown.

CHAPTER 9: Fireplaces, Entry Gates, and Mailboxes

Stone has long been a preferred building source because of its strength and durability. Fireplaces are some of the most requested amenities in homes, and with the advances in premanufactured fireboxes and one-piece chimney flues, you can easily install one in your home. Entry gates and stone gates add a touch of class to a house's entryway, and building a stone mailbox adds another appealing element in your landscape. This chapter will detail how to add a stone veneer to an existing unit with a **firebox**, which is a heat-tolerant insert added into an existing wall normally made of metal and firebrick, and also how to add a **chimney flue**, which is set inside the wall and through the roof for venting and to release smoke and cinder.

Installing a Stone Fireplace

MATERIALS YOU WILL NEED:

- Veneer stones, approximately 1 to 2 tons, flat on one side and textured on the other.
- Plastic tarp.
- Expanded metal lath.
- Masonry ties.
- Mortar — see materials list below:
    - Four bags of Portland cement.
    - One bag of lime.
    - One ton of sand.
- Access to water to mix mortar.
- Mortar pan.

Note: If you are building a hearth, you will also need flat stones for the top. If you are installing a stone mantel,

you will need at least 2-inch thick flat stones for the mantel.

TOOLS YOU WILL NEED:

- Large and small trowels
- Pointing tool
- Hammer and galvanized roofing nails
- Circular saw
- Stud finder
- Four wood two-by-fours, if installing a stone mantel, that are 8 feet long
- Level

Note: If existing trim has already been installed around the firebox, remove it and/or any glass screens.

INSTRUCTIONS:

1. Use the stud finder to locate wooden studs on which to attach the metal lath. Cover the floor around the fireplace with plastic tarp to prevent any damage to the flooring.
2. Measure out the dimensions around the fireplace, and cut your metal lath to fit inside the dimensions. You can easily cut the metal lath using tin snips, which are scissors meant to cut tin, but you should wear gloves to prevent cutting your hands and fingers. Locate the wood studs behind drywall by tapping lightly along the base of the wall. When you hear a solid noise, mark the area with a small "x."

3. Nail the pieces of metal lath to the studs in the wall using galvanized nails.
4. If you will use natural stone and want to extend the fireplace out into the room and not flush against the wall, you will need to build a hearth, which is the base extension of the fireplace. If you have carpet, you will need to remove enough of your carpet to build a stone hearth on; if you have wood or tile, you will not remove anything.
5. Attach metal lath to the floor as far out as you want your hearth.
6. If you have purchased precut stones, begin arranging them how you want them placed on the metal lath around the firebox. Mix your mortar, and apply a thin coat of mortar to the metal lath on the wall and floor.
7. Insert masonry ties for about every other stone, and stagger them as you place them in the mortar. These masonry ties will add stability for the rocks to hold to.
8. You will want to completely cover the metal lath with mortar but not too thick. Allow the mortar to dry overnight.
9. If you will install a floor-to-ceiling fireplace surround, which is a firebox surrounded by stone, brick, or rock, decide where you want your mantel. You can leave a space for a large wooden mantel, which you can purchase cut to size and finished with stain or varnish,

or if you want stone, you will need to leave enough room to allow for the width of the stone you are using. It is wise not to have a fireplace mantel extend too far over the fireplace itself, as it will deter airflow and may be structurally unstable.

10. You will need to brace these stones with wooden two-by-fours when you adhere them to the fireplace to allow them to set properly.

11. Butter one corner stone with mortar, and apply it to the lath. **Buttering** stone means you apply mortar to the stone itself and then adhere it to the wall rather than applying mortar to the wall and adhering the stone. This prevents too much mortar on the wall and ensures a good coating of mortar to each stone.

12. Apply pressure by holding the stone in place for a few seconds, and then wiggle it in place. Make sure you take your time and work slowly.

13. Continue with each stone until they are all in place. You may have to cut stones as you go to fit properly. You should place the stones as close together as possible without completely touching. Most fireplaces that use stone or rock do not have a specifically grouted surface, and the mortar acts as a grout joint.

14. If the mortar looks as though it is drying too quickly, use a spray bottle to mist it with water. Build out your hearth at the same time, and lay flat stones, such as

flagstone or slate, on the top. Make sure the hearth is as level as possible.

15. After all the stones are in place, you will want to trowel in mortar between all the stone joints.

16. After an hour, take your pointing tool, and remove any excess mortar. Use a whisk broom to brush the stones and remove any additional mortar and texturize the grout joints. Texturizing with the broom gives the mortar joints a rough surface. Make sure to remove any mortar from the face of the stones before it hardens and becomes hard to remove.

17. After a couple of days, the mortar will be cured and dried. Use a damp towel to clean the haze, which is a dusty film from the dried mortar, from the stones. You may want to use a clear sealer, but it is not necessary. The sealer will help keep the stones and mortar joints from getting dirty.

18. Wait at least two more days before you use the fireplace so you do not crack the mortar with heat. If the mortar does crack, you will have to chip out the mortar and reapply. If the stones become loose, you may have to rebuild the entire structure.

Depending on the mantel you have chosen, such as wood or stone, you will need to install the mantel while you build the interior chimney. If you have chosen wood, you can use wood glue and nails to attach the mantel to the lath. If you have chosen to use flat stones, such as flagstone, make sure the mantel does not extend out from the wall any more than 8 inches, with 2 inches being the depth it will be set back into the fireplace. A mantel that sticks out farther than 6 inches will need a brace for structural integrity. You will need to install masonry ties underneath the stones to help hold the stone in place. Butter the stone, and set in place the same way as the surrounding stones, only use the wooden two-by-fours cut to the right height to brace the stone while it sets. Leave the stone to cure and dry for two days. Let it dry for another two days before you place anything on the mantel.

Installing an Entry Gate

Entry gates add a certain air of elegance to the front of any driveway. Just walking through a stone entryway adds an element of mystery and wonder. The following section includes instructions on how to build three kinds of stone entryways: basic pillars, pillars with a gate, and an entry with an arched entry – which is also called a moon gate. Building the latter requires advanced masonry skills, which have been covered in other chapters of this book, specifically those discussing mortared arches requiring wooden forms.

Basic stone pillars

The simplest entry gates to build are stone pillars, or piers as they are called. **Piers** are shafts of layered rock that have an attached wall if they are used as entry gates.

Piers are seen from all sides so you should use aesthetically appealing stones in this project. The stones should also be proportioned for durability and for visual effect. For instance, a 6-foot tall pier should be at least 30 inches square and needs to have a foundation that extends 1 foot on all sides. You may also want to have a bead near the top of the pier, which is a small ledge that caps the top. If you can avoid using veneer stone and instead build the pier from solid stone, you will have a more durable pier better capable of withstanding weather elements. Veneer stone may pop off during cold weather. You can either dry stack or mortar your stones, but for the purposes of this project, mortared pillars will be explained.

MATERIALS YOU WILL NEED (FOR BASIC STONE PILLARS THAT ARE 6 FEET TALL AND 36 INCHES SQUARE):

- Three tons of stone with flat surfaces and square corners.
- Nine bags of Portland cement.
- One cubic yard of gravel.
- Two-thirds cubic yard of sand.
- Masonry ties.
- Mortar made from:
    - Six bags of Portland cement.
    - Two bags of lime.
    - One and a half tons of sand.
    - Access to water to mix mortar.
    - Mortar pan.

TOOLS YOU WILL NEED:

- Wheelbarrow
- Hoe
- Pick and shovel
- Trowels, large and small
- Pointing tool
- Wire brush
- Stone chisel
- Mason's hammer
- Level

INSTRUCTIONS:

1. Dig two 48-inch square trenches for the footings on both pillars. You should dig these below the frost line, and they should not contain any roots or debris. **Note: You can find your frost line by checking with your local building inspector.**

2. Start mixing your cement using two shovels of cement, four of sand, and six of gravel, plus water.

3. Pour the foundation, leaving approximately 2 inches of earth above the foundation's top.

4. Begin mixing your mortar. Lay the most rectangular or square stones around the outside dimensions of the pillar, and fill in the center with odd-shaped stones, large riprap, and mortar.

5. Keep applying the stones, buttering the base and not each individual stone with mortar. As you build up, make sure you overlap your stones for strength.

6. If you want a bead, or ledge, around the top of your pillar, place larger, flatter stones offset from the base, and mortar these stones liberally.

7. Place one more level of stones on top of the bead to help keep it in place.

8. Keep the pillar wet for two days to cure the cement, and then let it dry.

Building entry gates

Entry gates are stone pillars with swinging gates and are a little more difficult to build. If you plan to install a gate in your pillar, you must plan for it in advance so that you can embed the hardware necessary to hold the gate in place. If you install a gate in an existing wall, you will need to disassemble part of the wall to include the new pillar for strength. Simply putting in anchor bolts, which are long bolts that come out the other side of a structure and have a nut on one end to tighten it in place, will not hold a gate for long and will crack the stone pillar eventually.

Gate with stone wall and pillar

MATERIALS YOU WILL NEED (ENTRY PILLARS, WITH STONE WALL ATTACHED AND SWINGING GATE):

- Three tons of stone, up to 36 inches long, with flat surfaces and square corners.
- Nine bags of Portland cement.

- One ton of gravel.
- Two-thirds cubic yard of sand.
- Gate, gate hinges, pintles, and latch hardware.
- Mortar — see materials list below:
    - Six bags of Portland cement.
    - Two bags of lime.
    - One and a half tons of sand.
    - Access to water to mix mortar.
    - Mortar pan.

TOOLS YOU WILL NEED:

- Wheelbarrow
- Hoe
- Pick and shovel
- Trowels, large and small
- Pointing tool
- Wire brush
- Stone chisel
- Mason's hammer
- Level

For a pillar to hold a gate and to attach to a wall, you will have to build the hardware into the pillar itself during construction, and tie the pillar to the wall with tie stones. *For more information on tie stones, see Chapter 6.*

INSTRUCTIONS:

1. Dig two 48-inch square trenches for the footings on both pillars. You should dig these below the frost line, and they should not contain any roots or debris. **Note: You can find your frost line by checking with your local building inspector.**

2. Start mixing your cement using two shovels of cement, four of sand, and six of gravel, plus water. At the same time, you will need to dig foundation trenches for the walls that you will attach to both pillars.

3. Pour the foundations, leaving approximately 2 inches of earth remaining above the foundation tops.

4. Begin mixing your mortar.

5. Lay the most rectangular or square stones around the outside dimensions of the pillar, and fill in the center with odd-shaped stones, large riprap, and mortar.

6. As you build your pillar, start building the courses of your walls as well, integrating the wall with the pillar using overlapping stones.

7. Keep applying the stones, buttering the base and not each individual stone with mortar. As you build up, make sure you overlap your stones for strength.

8. The hardware for the gate embeds in the stonework as you build; therefore, when you get to the proper height for the base of your gate, you will need to mortar in pintles. **Pintles** are simply pins or bolts on which another part pivots. Pintles used in gates are L-shaped, flat pieces that have a bolt facing upward on which a gate hinge mounts.

9. Mortar the wall liberally, and lay your stones directly on top of the flat part of the pintle.

10. You will also want to embed your latch mechanism on the opposite pillar as you build it.
11. If you want a bead, or ledge, around the top of your pillar, place larger, flatter stones offset from the base, and mortar liberally.
12. Place one more level of stones on top of the bead to help keep it in place.
13. Keep the pillar wet for two days to cure the cement, and then let it dry.

Arched entries and moon gates

An arched entryway that has a stone veneer wall and stone steps.

Moon gates.

During the mid to late 1800s in Europe and America, people built high arches into entry gates because carriages made their way through them. The higher the arch, the more prestigious people perceived the home because some carriages had higher wheels and seats on which the carriage driver perched, some of which were 10 to 12 feet tall. Homes needed high arches so the driver could pass underneath them. Pillars for high arches need to be a minimum of 12 feet apart to allow any car or truck to pass through without any danger of running into the pillars.

Building a moon gate is a complicated procedure and will take all the skills learned in mortaring walls, pillars, and arches — dry stacking is never recommended for arches. Moon gates were designed for a footpath or horse trail but not for vehicular traffic. It can be as little as 8 feet tall and is usually built into a stone wall.

Moon gate or arched entry with a stone wall

MATERIALS YOU WILL NEED:

- One ton of stone, up to 12 inches long, with flat surfaces and square corners — this amount of stone will yield 18 cubic feet of mortared wall.
- Nine bags of Portland cement.
- One ton of gravel.
- One ton of sand.
- Gate, gate hinges, pintles, and latch hardware.
- Mortar — see materials list below for every 36 to 45 cubic feet of wall:
    - Three bags of Portland cement.
    - One bag of lime.
    - One ton of sand.
    - Access to water to mix mortar.
    - Mortar pan.
- Arch form.
- Two sheets of ½-inch construction-grade plywood.
- Several lengths of two-by-fours to use as temporary braces to hold the arch in place.

TOOLS YOU WILL NEED:

- Wheelbarrow
- Hoe
- Pick and shovel
- Trowels, large and small
- Pointing tool
- Wire brush
- Stone chisel
- Mason's hammer
- Claw hammer
- Nails
- Jigsaw
- Level

NOTE: This project assumes you have either partially built or completed one or two courses of a wall that you are building your arched entry or moon gate into.

INSTRUCTIONS:

1. Cut your plywood with the jigsaw into a semicircular shape. Use the one cut as an identical template for the other. Nail small spacer blocks approximately 2 inches in between both semicircular arches, and place over the course of the wall, bracing in place with the two-by-fours.

2. You should shape stones to taper into the arch when mortared. You will mortar the stones on the arch with the same approximate mortar joint as the wall.

3. Build the lower half of the moon gate by mortaring in your stones against the wooden arch as you complete building the courses of your wall. Use small stones as

wedges as necessary to keep the lower arch level and tight. Let the base cure for two days.

4. Remove the semicircular wooden arch and bracing boards. Point up the spaces in the lower arch by troweling on mortar and getting it into all the cracks between stones.

5. Clean all excess mortar, and let it cure and dry for two days.

6. For the top half of the arched moon gate, set the semicircular form on top of the stones in the bottom arch and wall, and brace with two two-by-fours in the center, on top of the arch, as well as along the sides of the arch.

7. Coat each surface of stone with mortar, and set in place on the top form.

8. Remove the form after two days.

9. Use the trowel and pointing tools to fill in all the cracks with mortar.

10. Keep damp for two days to cure and dry.

Building Stone Mailboxes

Mailboxes made from stone add a stylish addition to the front of any home. You can incorporate the mailbox, newspaper tubes, and even a locking or keyless entry box where you can safely and securely store extra house keys for service people or pet sitters.

MATERIALS YOU WILL NEED:

- Half ton of stone, up to 12 inches long, with flat surfaces and square corners.
- Four bags of Portland cement.
- Half ton of gravel.
- Half ton of sand.
- Metal or plastic mailbox insert, without post, keyless or locking lock-box, and newspaper delivery tube (optional).
- Mortar — see materials list below:
    - Two bags of Portland cement.
    - One bag of lime.
    - Half ton of sand.
    - Access to water to mix mortar.
- Mailbox frame.
- Four pieces of two-by-fours cut to the desired height of your mailbox, adding 2 additional feet to the length to allow the mailbox to sink securely into the ground.
- Several lengths of two-by-fours to use as braces to hold the frame together.

TOOLS YOU WILL NEED:

- Wheelbarrow
- Mortar pan
- Hoe
- Pick and shovel
- Trowels, large and small
- Pointing tool
- Wire brush
- Stone chisel
- Mason's hammer
- Claw hammer
- Nails
- Jigsaw
- Level

INSTRUCTIONS:

1. Dig a foundation trench approximately 2 feet deep and to the desired width and length of your planting box. You have to make this foundation level or the mailbox will lean.
2. At the top of the trench, set a frame for your mailbox made from four pieces of two-by-four, all braced with two-by-fours cut to length and nailed together. Make sure the pedestal frame is even when you place it inside the foundation trench by using a level.
3. Mix the cement — *as instructed in Chapter 3.* Pour the mixture, which should be the consistency of thick pancake batter, into the foundation trench until it fills it within 2 inches of the top.
4. Drag a two-by-four across the frame in sawing motions to clear away any excess cement and level your base.

5. Tamp, which is to tap lightly in succession, the cement down with a rake until all the rock in the mix settles. .

6. After the cement begins to thicken, typically 15 minutes to half an hour, skim the cement with a cement trowel until the surface is smooth and uniform. Allow one day for the cement to harden before laying your stone.

7. Use a masonry trowel to set a line of mortar around the perimeter of your foundation, and set your stone into it.

8. Keep your mailbox level by tapping the stones as you lay each course. Remember to overlap all your corners and interior stones for strength.

9. If you plan to add a newspaper tube or lock box, you will need to add them as you lay the stone so you can embed each item. Typically, a keyless lock box is below a newspaper tube, and both are below the mailbox. The United States Postal Service has strict regulations as to the required height of mailboxes so check these requirements before placing your mailbox in the stone structure at **www.usps.com**.

10. When you reach the desired height for the lockbox or newspaper tube, place them on the courses of stone with the openings to the outside, and mortar into the courses as you go. When you have reached the required height for the mailbox itself, lay the mailbox, with the door opening outward, onto the stone course.

11. Fill in around the mailbox with stone, as you did the newspaper tube and/or lockbox.

12. You can add a flat top by laying a large flat capstone at the top of your mailbox pedestal.

13. Carefully clean any remaining mortar from the stones, and let dry/cure for two days.

Building a Brick Planter Around an Existing Mailbox

If you already have an existing mailbox or stone mailbox that you like, you can add a planter around the base for flowers using a few bricks and some mortar.

MATERIALS YOU WILL NEED:

- Half ton of stone or brick with flat surfaces and square corners.
- Mortar — see materials list below:
    - Two bags of Portland cement.
    - One bag of lime.
    - Half ton of sand.
    - Access to water to mix mortar.

TOOLS YOU WILL NEED:

- Wheelbarrow
- Mortar pan
- Hoe
- Pick and shovel
- Trowels, large and small
- Pointing tool
- Wire brush
- Stone chisel
- Mason's hammer
- Level

INSTRUCTIONS:

1. Before you begin building your planter, ensure your ground is level. A brick planter does not require a foundation, but the ground does need leveling prior to laying your stone.

2. Lay your stones or bricks out to determine how big you want your planter. You can leave the bricks down since you do not need to mortar them to a foundation.

3. Mix the mortar, and trowel it on the foundation brick. Spread it out, and into the foundation brick joints, and add brick, working from the corners out. Remember to

overlap all your brick joints and especially the corners to strengthen the planter.

4. Press the brick down until there is approximately ¾ inch of mortar between your bricks, and use your level to determine if you are even on all sides as you go.
5. Continue building the sides until they are as high as you want them.
6. Level them all once more, clean off any excess mortar, and let it cure for two days, keeping it moist and then allowing it to dry.
7. Fill with good potting soil and add plantings.

If you have existing stone structures already in place in your landscape and they need some sort of repair, you can easily make these repairs yourself, but you should leave some restoration work to professionals.

CHAPTER 10: Restoring Existing Stone Structures

If you have existing stonework, whether a sagging foundation or a crumbling porch post, it is sometimes much easier to repair the stonework rather than tear it down and create a new one in its place. In some cases, it is simply a matter of replacing mortar, especially in the cases of homes built in the 1800s with a lime and sand mortar. In northern climates where basements are prevalent, it is necessary to keep the basement foundation strong because it is often the foundation of the home or building itself. If it begins to deteriorate, the entire living structure may cave in. One of the most important rules to follow in good restoration work is that the restoration should duplicate the existing work.

Structural Foundations

Most early foundation work did not have the structural advantage of concrete for stability, and stones were simply set down into the earth without footings. These bases, although set on the largest and flattest stones available, were usually uneven in minute ways, which caused eventual cracks that ran up the walls over the course of the years.

Structural engineers now work on every major structure to calculate how much a structure will settle and ensure the total weight load is in proportion with the building's expected degree of stability. Most stone restoration work now done to foundations requires removing concrete and stone in small sections at a time, which is a laborious, but essential, process in stabilizing existing structures.

Depending on the foundation structure, you may want to hire an outside company that specializes in foundation restoration work. Many times, what you think is a simple restoration job can turn in to having to actually move the structure itself to a certain degree; building trusses to hold up the structure as the foundation is replaced or repaired; digging and pouring concrete footings to support piers for floor beams; and moving huge stones in place, which requires heavy machinery and a complicated series of posts and pulleys. It is always wise to seek professional restoration companies to come out and give an estimate of costs involved in restoring any type of structure.

But, if a stone foundation simply needs new mortar, you can easily do this by mixing mortar and applying it using a small trowel and pointing tool. You can brace foundations using wooden beams and frames, which will add stability.

Dry Stone Wall Restoration

The biggest problem with dry stone walls comes from roots and vines causing underlying foundation instability, which causes the stones to cave in and fall. These walls fall because of a crumbling base, whereas dry stone retaining walls fall from the top as a result of uneven pressure of earth or water. In historical restoration of dry stone walls and buildings, each stone is picked up off the wall, numbered, and when the foundation is repaired, replaced exactly in the same position it was located before the foundation restoration. If you are interested in doing your own foundation restoration, you should do this as well.

When you restore a dry stone wall, you should rebuild all courses one course at a time, ensuring you pay special attention to replacing the missing stones with stones that will fit the original look of the wall. The work you do should not be a temporary remedy; if you are going to restore a wall, make sure you restore it properly.

Mortared Wall Restoration

Due to the nature and durability of mortared walls, it is much harder to repair a mortared wall or structure than it is to restore a dry stacked wall. You will need to dismantle the damaged sections, and you will need to pour the footing again if it has broken. You will need to chip off any old mortar, which will not be an easy task if it was made from Portland cement. The Portland cement, a watertight mixture, creates such a strong adhesive bond that when you chip the cement off the brick or stone, you may crack the stone itself. If the structure was made with the old lime-and-sand mortar, you can easily remove it because this mortar is soft and crumbles with age.

Another issue with repairing mortared structures is matching the mortar mix. If you are repairing a historical structure and have to adhere to strict guidelines from a historical society, they will most likely require you to have the mortar analyzed by an architectural organization or building conservation group. Look up this information in your own area by starting with your own city building authorities.

You should complete repairs to historic buildings using materials and techniques that match the ones originally used as much as possible. The matching materials will age in a similar way as the originals did, and you can limit the possibility the restoration will harm the old building brick and stone. Using Portland cement on old building stone quickly deteriorates the fabric of the stone at a quicker rate than more traditional mortar mixes. Historical societies only discourage using newer cement fabrications if it will result in the degradation of the original stone.

You may also want to incorporate natural materials to "dye" the mortar to match the existing mortar. Several commercial dyes available will add the look of age to the mortar, but you can also use chimney soot, dirt, and local river sand to color the mortar.

Pointing

Pointing is a relatively easy process, although very few people specialize in it anymore because of the time involved. The process of pointing involves picking out old mortar on buildings with a narrow chisel or pointing tool and pushing in new mortar. Most pointing is done on historical buildings because of the cost of the laborer's time – it can take more than four weeks to do one side of a building. Pointers charge a flat fee based on the footprint of the building, and it can range as little as $2,000 to more than $250,000, depending on the building and the restoration required on the mortar. They can easily remove old lime-and-sand mortar because it comes out easily, but when someone used Portland cement and various quick fixes to poorly repair broken brick or mortar, they cannot replace them.

When you do your own pointing work, you need to dig into the old mortar joints about 2 inches, wet the stone with a spray bottle of water, and push in the new mortar carefully. You must immediately clean the stone or brick of any excess mortar to prevent the stone or brick from discoloring. Keep the mortar recessed into the joints, and keep each section you work on damp for two days so that the moisture will not draw out of the surrounding stones and weaken the structure. Laying moistened burlap or plastic tarp over each section after it has been pointed and misted is a good way to keep it damp until it cures.

Recycling Old Stone and Brick

Stone and brick are perfect building materials to recycle because most of the time you can reuse them all — even stone veneer. Brick from old streets contains a high amount of metal ores, making the brick almost indestructible, and many private house builders loaded their trucks with these old street bricks to use in fireplaces and foundation work. Even the dockside streets of Savannah, Georgia are lined with recycled stone: Ballast cobblestones used to weigh down sailing vessels from the late 1700s and 1800s were used to make the streets along the river. Cobblestones are great for visual appeal but are pretty hard on vehicular and human traffic because they are uneven and rounded and are quite slick when wet.

People always tear down old brick chimneys to make way for new chimneys that feature newer and longer-lasting brick. People now seek out these old bricks for their unique discolorations and use them as veneer in lofts and cabins. Even broken concrete from shopping center parking lots is used as filler on construction sites because the mixture of the broken concrete and sand acts to percolate the soil in drainage ponds. **Percolation** is defined as the ability for water to sift and filter through soils rather than puddle and pool up, causing drainage issues. The more the soil is aerated through the soil additives, the more likely it will drain properly. People have also used broken concrete from their driveways to edge flower beds and create mosaic patios in their backyards.

When it is not feasible to repair or restore something built with stone, you can always call a salvager who will come to your home to assess the value of your stones and possibly haul off the stone and brick for little cost.

Rather than replace the entire driveway, the homeowners cut out the areas with large cracks and used brick pavers to replace those sections. Photo courtesy of Douglas R. Brown.

CHAPTER 11:

Caring for Stone

Stone and brick require very little care, but you must do certain things to ensure that all of your hard work will not be in vain. Keeping brick clean from smoke and soot, as well as cleaning moss and lichen off stone when you do not want it growing there, are topics covered in this chapter.

Brick

Caring for brick, both interior brick and exterior brick, is not terribly difficult. Unless you have soot buildup on your fireplace front, you really do not have to clean brick once properly installed. But, there are instances when mosses, lichens, or mold may grow on the outside of your brick, and you may not want them there. Also, with time, brick will absorb dirt and dust and may need an occasional general cleaning. Bricks in damp basements also tend to become a breeding ground for mold and mildew so you will need to treat the brick if molds begin to appear.

Bucket and brush method

This is the most widely used method to clean brick. You can use this method on all colors and textures of brick, but you must choose the correct cleaning solution. Simple bleach and water may not clean as well as a specific muriatic acid solution would. A 10 percent solution made from muriatic acid — one part muriatic acid and nine parts water — will clean most brick walls. Please keep in mind that this solution can be dangerous to use so you must take all safety precautions when using it. You should wear gloves, protective goggles, a face mask, long pants, and shirts with sleeves when using this solution, and never mix it stronger than 10 percent. Adding more acid will not make it clean faster and may damage the brick.

SUPPLIES NEEDED:

- Muriatic Acid
- Water
- Safety Goggles
- Face mask or respirator
- Nylon bristle brush
- Gloves, preferably ones that protect from chemical agents
- Bucket

INSTRUCTIONS:

1. Make sure the mortar on your project has had the chance to harden for three to seven days. Remove all large mortar particles with hand tools such as chisels or trowels.

2. Saturate the brick wall with clean water — if you are working outside, use a hose. If you do not saturate the wall properly, making sure it has soaked up plenty of water, you may create visible cleaning solution stains. Keep the brick saturated until you are ready to clean, and work in shade or early morning hours when the brick will not dry too quickly.

3. Use a stiff-bristled nylon brush, preferably one that you can attach to an extension pole for higher surfaces. Dip the brush into the cleaning solution, and apply to the wall, scrubbing briskly. Scrub the brick, not the mortar joints. The mortar is more porous and less compacted than fired brick and may crumble if continuously scrubbed. Leave this solution on the brick for three to six minutes. When you have finished with an area, rinse thoroughly with water.

High-pressure water cleaning

This is an ideal method to use if you have access to a high-pressure cleaning system and if the surfaces you are cleaning are tall walls, outdoor patios, and fireplaces. Using a high-pressure hose causes a great deal of overspray so do not use it for indoor applications. The pounds per square inch (PSI) is between 400 and 800 PSI for home-use units and is as high as 1,000 PSI for commercial units. The home-use pressure cleaning units are sufficient.

SUPPLIES NEEDED:

- Muriatic acid
- Water
- High-pressure washer
- Garden sprayer, low pressure
- Safety goggles
- Face mask or respirator
- Gloves, preferably ones that protect from chemical agents
- Bucket

INSTRUCTIONS:

1. Wait until all mortar has had time to dry and harden — at least seven days. Clean any hardened mortar off the brick with hand tools, such as chisels or trowels. Protect any other surfaces that you could damage by the use of high-pressure water.
2. Saturate the area you want to clean with water. You can use a common water hose for this procedure. When you completely saturate the wall, apply the muriatic cleaning solution — use a 10 percent solution of muriatic acid, which is one part acid to nine parts water. The cleaning solution is best applied using a standard low-pressure sprayer, such as the kind used for pest control. Allow the cleaning solution to remain on the brick approximately three to six minutes.
3. Rinse the brick with the high-pressure water, starting from the top and making your way to the base. The best high-pressure tip to use is the fan type, preferably stainless steel. The fan-type tip will allow more water

to spread over more surface, making the job require less time and effort. You should opt for stainless steel because it does not rust.

Sandblasting

Sandblasting is a controversial method of cleaning brick because it will erode some older brick and mortar joints, and, therefore, persons wanting to maintain the historical integrity of old brick structures shun it. But, it is the most efficient and cost-effective method of removing almost anything from the brick's surface, including paint. Sandblasting is not recommended by any casual user, and should you decide to have your structure sandblasted as your preferred method for cleaning your brick, you should contact a qualified professional.

The process of sandblasting entails using a powerful compressor and a blasting tank filled with abrasive materials, such as finely graded granite, sand, or quartz that all have sharp cutting qualities, and using air pressure to apply the abrasive materials to the surface. There are two types of gradation, A and B, to use on brick. The type A gradation is used to lightly clean brick, and type B is used where there is heavy staining or painting and more abrasion is needed.

Because commercial companies are the best source to use to complete this type of brick cleaning and because you can hurt yourself or others, this book does not include the steps for sandblasting your brick. All sandblast professionals require Occupational Safety and Health Administration (OSHA) certification, and this endeavor requires large, expensive equipment.

Cleaning fireplace brick

You can clean both the interior and exterior brick on your fireplace by one of two methods: wet or dry. Both of these methods will clean smoke and soot. If your fireplace is not surrounded by a hearth and the fireplace is within close proximity to carpet, you may prefer using the dry method.

SUPPLIES NEEDED FOR WET CLEANING:

- Dishwashing detergent — any kind
- Water
- Safety goggles
- Face mask or respirator
- Nylon bristle brush
- Gloves, preferably ones that protect from chemical agents
- Bucket
- Vacuum

INSTRUCTIONS FOR WET CLEANING:

1. Clean all debris from the interior of your fireplace. Once you clean the interior, vacuum lightly to remove all light dust and loose mortar.
2. Combine a tablespoon of dishwashing detergent into a bucket of water. Using a stiff bristle brush, preferably nylon, dip the brush into the bucket, and scrub until

you remove the stains. Soak up any excess water with a sponge, and continue until you clean the bricks.

SUPPLIES NEEDED FOR DRY CLEANING:

- Dishwashing detergent — any kind
- Table salt
- Safety goggles
- Face mask or respirator
- Nylon bristle brush
- Gloves, preferably ones that protect from chemical agents
- Bucket
- Vacuum

DRY CLEANING INSTRUCTIONS:

1. Clean all debris from the interior of your fireplace. Once you clean the interior, vacuum lightly to remove all light dust and loose mortar.

2. Combine 1 ounce of dishwashing detergent and 1 ounce of salt together, and work it into a paste. Apply this to the brick, and wait approximately ten minutes. Scrub with a stiff bristle brush until you remove the stains.

Cleaning mold and moss off brick

Some people prefer to keep mosses and ivy growing on their brick, and some do not. In any case, you do not want mold on your brick because it shows the brick is retaining moisture, which will cause the brick to soften and crumble. Mold also produces digestive enzymes that eat the brick, which can also cause structural damage. Inhaling mold spores can lead to health hazards, such as asthma, sinusitis, nausea, rashes, and various other allergies.

Cleaning mold and mildew off brick

SUPPLIES NEEDED:

- Chlorine bleach
- Water
- Safety goggles
- Face mask or respirator
- Nylon bristle brush
- Gloves, preferably ones that protect from chemical agents
- Bucket

INSTRUCTIONS

1. To protect yourself against the mold, you should wear gloves, protective eyewear, a face mask, long pants, and a shirt with sleeves. You can find mold by visually spotting it or smelling it — mold has a distinctive aroma of musk.

2. The most favored cleaning solution for mold and mildew on brick is common chlorine bleach. The standard dilution is to use equal parts of bleach and water. Using a stiff bristle brush, preferably nylon, dip the brush into the bucket of solution, and scrub it until you remove the mold. Leave the bleach on the bricks for approximately three minutes.

3. Rinse the bricks thoroughly with clean, clear water, and then let it dry. If the mold or mildew reappears,

you may have a moisture retention issue and will need to reapply the bleach solution.

If you have moss growing on your brick and it is unsightly or unwanted, you need to clean it off the brick and need to prevent it from coming back. Moss is very resilient and will grow back if not removed effectively. Gloves, safety goggles, and a face mask are recommended because you will once again work with chlorine bleach, which is harmful to skin, and you will spray it onto the wall.

Cleaning moss off brick

SUPPLIES NEEDED:

- Chlorine bleach or distilled white vinegar – *see notes*
- Water
- Safety goggles
- Face mask or respirator
- Gloves, preferably ones that protect from chemical agents
- Nylon bristle brush
- Bucket
- Broom

INSTRUCTIONS:

1. Use your broom to remove as much of the surface moss as possible. This will allow the bleach solution to more effectively get to the source of the moss.
2. Spray the bleach solution onto the brick. Be very generous when spraying the brick. Leave the solution on for three minutes.

3. Go back over the bricks with the broom to remove any additional moss that has come loose. Rinse the bricks thoroughly with water.

Note: If you do not want to use bleach because you are either allergic to it, have animals around that may become affected, or you have plants in close proximity to the brick that you will clean, you can substitute distilled white vinegar. It is not as quick or effective as bleach, but it is less harmful to the surrounding landscape. You may have to treat the brick again for moss removal.

CHAPTER 12:
Using Rock and Stone for Crafting and Art

While rock and stone are beautiful just as they are, you can craft many projects using them to create many clever and interesting home décor items. This chapter includes just a few of the many things you can make from rock and stone.

CASE STUDY

Lin Wellford, artist and author
ArtStone Press
9328 Highway 62 E.
Green Forest, AR 72638
www.linwellford.com
linwell@windstream.net

Lin Wellford wrote more than nine books on painting three-dimensional art on stones and rocks, and she is a phenomenal artist who can make a simple river stone into a flower garden, a sleeping fawn, or even a small country mansion.

Wellford's first encounter with the beauty of stones and using these stones to create art came in 1978. "I grew up in a rockless place (Sarasota, Florida), so rocks and stones intrigued me when we moved to the Ozarks in 1978. I picked up rocks along the creek while my children were swimming and loved the shapes and the natural feel of them. One looked so much like a little bunny that I decided to take it home and paint on a few details. The results were so amazing. Three-dimensional art. I was hooked," Wellford said. Wellford said that she "prefers to choose rocks that are fairly smooth without being slick" because it is hard to keep paint on shiny rocks, and rough rocks make it hard to paint details.

After many more years of painting hundreds of animals, in 1994 she wrote a book, *The Art of Painting Animals on Rocks*, in which she shared her unique talent for painting rocks, as well as techniques and tips on how to choose rocks to match a particular subject.

Wellford's first book became a best seller and spawned many more on many other painting subjects, such as painting flowers and little cottages, painting rock pets, painting projects for children and beginning painters, and a book devoted to painting manufactured stones and edging to turn any outdoor space into a garden gallery.

Wellford has appeared on dozens of local and national televisions shows, including HGTV's *Carol Duvall Show* and published numerous articles in national and regional magazines.

Wellford has taught dozens of classes and workshops over the years and urges people who are interested in learning rock painting to join the free Yahoo!® group, **www.groups.yahoo.com/group/rockpainting**, for patterns, advice, and lots of encouragement from other members.

Poetry pebbles

MATERIALS YOU WILL NEED:

- One small bag of river rock, polished or unpolished.
- Two or three small — 18/0, 10/0, or 5/0 — round paintbrushes.
- Container for water.
- Acrylic paints, formulated for outdoor use.
- Clear spray polyurethane.

INSTRUCTIONS:

1. Clean your river pebbles well with soap and water. Dry thoroughly with towels. Creating poetry pebbles means you will need to have a lot of connecting words, such as "and" and "with." You will also need emotion words, such as "happy" and "laughter."

2. Paint these words directly on your pebbles. Add some exclamation points or commas if you like. Use your imagination to come up with more words.

3. Let the rocks dry about five hours, and spray each stone with the clear polyurethane to help seal the acrylic onto the stone. Take them to the beach or park to have fun "writing" your poetry with the pebbles. Be sure to take a photograph.

Words to use for the poetry pebbles:

she	he	to	can	will	love
dog	walk	happy	sad	should	could
in	out	your	me	hand	heart
a	my	dance	thoughts	simple	hold

This craft builds the creative process in the brain and helps children learn words and sentence-building skills. You can also play a game with friends by letting each pick ten pebbles from a box, without looking at them, and challenge them to build a poem from the words they chose.

Rock paperweight

A rock or stone, all by itself, makes a wonderful paperweight, but you can decorate an ordinary stone into a decoupaged paperweight that carries a message. These make great gifts, and they are fun and easy to make.

MATERIALS NEEDED:

- One or more smooth stones, at least 3 to 4 inches wide.
- Thin wrapping or crepe paper in different colors.
- One square of felt.
- Decoupage glue — available at craft stores.
- Paintbrush — angled or flat work best.
- Brown paper bag.
- Wax paper.
- Clear spray polyurethane.
- Craft glue that dries clear.
- Scissors.

INSTRUCTIONS:

1. Tear your crepe paper in small strips.
2. Lay out the paper bag, and then cover it with wax paper. This creates your work surface, and the wax paper will prevent the paper and glue from sticking together when the glue dries.
3. Cover a small area of your stone with decoupage glue, and place the crepe paper over it, applying more glue on the top of the paper.
4. Keep adding strips of paper and glue until you have built up the paper all over the stone.

5. Let your stone dry overnight. Once it has dried, spray the stone with the clear polyurethane, and let it dry for at least two hours.

6. Cut a small oval or round piece of felt, and attach it to the base of the stone.

Pet rocks

Pet rocks were popular in the early 1970s. They were sold in small boxes, much like the boxes used to carry home live pets, such as hamsters and turtles. You can make your own pet rocks by adding a few details, and give them to all your friends.

MATERIALS NEEDED:

- One or more smooth stones, at least 3 to 4 inches wide.
- Different colors of felt.
- Googly eyes.
- Yarn.
- Dimensional, such as puff, paint.
- Small pom-poms.
- Beads or other embellishments.
- Acrylic paints — black, blue, pink, white, and red.
- Brown paper bag.
- Clear-drying craft glue.
- Scissors.

INSTRUCTIONS:

1. Wash and dry your stones thoroughly.
2. Use the paper bag to build your pet rock on — this will protect tables from glue and help you to organize all your craft supplies.
3. Next, add the eyes. You can glue the googly eyes on with the craft glue or draw them on with the acrylic paints. You can add eyelashes and eyebrows next.
4. Add the nose by drawing it on or gluing on a small pom-pom. Add a mouth by painting it on with the dimensional paint.
5. Add hair or moustaches by using yarn or felt.

Rock spiders

This is a great craft for Halloween, and it looks pretty realistic when you have finished it. You can also make rock ladybugs if you prefer.

MATERIALS NEEDED:

- One or more smooth stones, at least 3 to 4 inches wide.
- Googly eyes.
- Black pipe cleaners.
- Hot glue gun.
- Glue sticks.

- Acrylic paints in black, yellow, and white — add red if you are making a ladybug.

INSTRUCTIONS:

1. Clean your stones with soap and water, and let them dry thoroughly.
2. Paint your entire stone with black acrylic paint, and let the stone dry for at least an hour.
3. Glue on the googly eyes with the hot glue, or paint your own eyes with the yellow and white paint.
4. Take your pipe cleaners and bend at a 90-degree angle.
5. Hot glue four ends of the pipe cleaners to the underside of your rock.
6. Hot glue an additional four bent pipe cleaners to the opposite underside of the rock.
7. If you like, add a smile to make it a friendly Halloween spider.

Variations of the rock spider include using red paint instead of the black to create a ladybug, covering each side to form ladybug wings. Add small dots of white to the red wings. Use the same pipe cleaners, but cut them down to 4 inches, and only use three pipe cleaners on each side. Bend at the same angle, and hot glue underneath the rock to make smaller legs.

Several books on the market show how to make clever crafts using stones as a painting surface, and the Bibliography contains these books.

Snowman doorstop

A simple keyhole paving brick makes an adorable snowman doorstop — and you can make variations to include making it into a turkey for Thanksgiving or a Santa Claus. You can even make it into a gingerbread man.

MATERIALS NEEDED:

- One keyhole paving brick — color does not matter.
- White enamel spray paint.
- Acrylic paints in black, orange, brown, and white.
- Black felt — one square.
- Scrap fabric for scarf.
- Hot glue gun.
- Paintbrush.

INSTRUCTIONS:

1. Lay your paving brick on a protected surface, and spray paint the front and sides.
2. Let it dry, and do the same with the back. It may take two coats if use a colored brick.

3. Paint on two rounded circles for eyes on the small square portion of the brick. Also paint in the mouth and the nose.

4. Add three to four larger rounded black circles for the coal buttons.

5. With the one square of felt, cut a large triangle that spans from one corner to the next. This makes the snowman's hat. Form a small rim by turning over the base of the triangle about a ½ inch. Hot glue the triangle around the head, securing the cut ends at the back.

6. Take a small piece of ribbon or scrap fabric, and tie onto the felt. Cut the felt tops to add fringe. Cut a long strip of fabric for the scarf, and fray it. Wrap around the neck area, and secure with hot glue. Cut the ends to add fringe.

Using Stone to Create Pebble Mosaics

Pebble mosaics, an enchanting art form, can incorporate hard surfaces into landscape design, called hardscaping. It is ideal for hot climates that would not otherwise support a lot of greenery, such as New Mexico or Arizona, and for climates where the cold robs your garden of color and plant life for a few months out of the year. The art of incorporating pebbles as mosaic hardscaping is Old World, having been used extensively in garden paths of Spain and castle entrances and garden areas of the United Kingdom. In the United States, you can find magnificent examples of pebble mosaics in the Waterworks Garden in Renton, Washington, and Lotusland, in Santa Barbara, California.

You may wonder what kind of pebbles you can use to create land art in your home. Hard, water-worn pebbles provide the best stones to use for these art projects. You can find them on beaches, rivers, and many quarries and stone yards carry a various display of pebbles. Long and thin pebbles work well for defining shapes. You can find flat pebbles for filling in larger surfaces, and you can use small pebbles, such as pea gravel, for detail. You can also incorporate tiles and glass marbles for interest. You can create pebble mosaics with or without using cement or grout. And, as all pebbles are comprised of different minerals depending on the region, they come in many different colors. This chapter contains a couple different projects you can use both inside and outside your home.

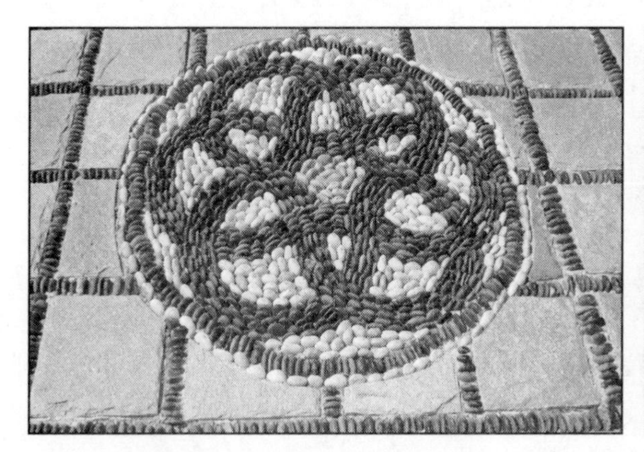

Learning the basics

Four basic rules for completing a successful pebble mosaic include:

1. Prepare a base.
2. Use side restraints.
3. Always set the pebbles vertically.
4. Pack the stones tightly.

When you prepare a foundation for a pebble mosaic, keep in mind that it must be solid. If the ground is soft, the pebbles will sink or get washed out by heavy rains. Clear your foundation of all grasses and roots, and tamp the ground well using the back of a flat shovel or walking firmly all over the base to pack the soil. You may want to use a garden hose to add water, wait for the water to absorb into the ground, and repack the soil even tighter.

When designing your pebble mosaic, you will need to add physical restraints to keep the pebbles from shifting. These restraints can be as simple as a form created from pressure-treated wood or an actual stone edging built on a concrete foundation. Using a form to create your pebble mosaic also makes it easier to see your pattern and to further design for landscaping.

When I did my first pebble mosaic, I laid them flat so that I could see more of the surface stone. But, in doing so, I lost quite a few stones when I grouted because there was just not enough of the stone embedded for it to stay put.

So I learned early on to always embed the stone vertically, never flat. It may seem like a more laborious process, but it will ensure your hard work does not loosen from the grout the first time you walk on it.

Pack your stones as tightly together as you can. You may even want to use a rubber mallet to pack the stones tight so they lock up against one another.

Types of stones and pebbles

The best types of stones and pebbles to use are water worn. This almost always ensures an intrinsic hardness and density of stone necessary for a pebble mosaic. The river rock sold in landscaping companies and home stores are perfect for this project, as are the smaller stones used in floral decorating. To reduce the cost associated with building larger mosaics, get stones and pebbles as cheaply as possible from stone yards and quarries, which may entail buying a larger quantity of stone to receive a cheaper price. Save the specialty stones for details. The best types of pebbles or stones to use are the granites, quartz, hard limestones, and fine-grained sandstones. Also choose stones for their color. Most commercial river rock will be gray in tones, almost black at times, and pea gravel is usually warm oranges, browns, and reds. Landscaping marble is white but is somewhat angular. You will use these colors the most.

The technique of laying stone

When working with pebble mosaics, you will learn through trial and error the best ways to create your own work. You can use wooden stencils laid out on the foundation and actually built around and on top of the stencil. You can also draw your design on sheets of white plastic, numbering each area with a specific color or stone to use, and lay this on the foundation, keeping it in place with blocks of wood. All of these methods are loosely defined as "in-situ," meaning "on-site." This is also called "the Mediterranean technique." You can also create what is known as a "pre-cast" mosaic. You create the pebble mosaic at a specific work site using concrete to set the stone, and transport it to the location you want to place it in.

Creating a small pebble mosaic stepping stone using "in-situ"

MATERIALS YOU WILL NEED:

- One large bag of pea gravel.
- Two large bags of river rock — landscaping stone.
- One bag of dry cement mix.
- One bag of sand — playground sand is fine.
- Approximately 15 bricks to use as the surrounding restraint.

TOOLS YOU WILL NEED:

- Garden hose (and access to water)
- Shovel
- Gloves
- One two-by-four, cut to approximately 2 feet long

INSTRUCTIONS:

1. Shovel out a rounded area where you want the stepping stone. You should shovel this approximately 4 inches deep and approximately 24 inches wide. Make sure the soil contains no debris, and it is packed tightly. Lay your restraint around the circle using the bricks. Be as neat as possible because this brick border will remain attached to the stone. You can fill in the empty triangular areas later. You should set the brick level to the top of the stepping stone.

2. Into this circle, add a mixture of two parts sand and one part dry cement — use gloves because cement can irritate to your skin. This mixture should be about 1 inch deep in the foundation.

3. Start setting your larger stones into the sand/cement mixture. Make sure to set them vertically and as tightly together as possible. Start from the center when working with a rounded object. Press down lightly as you set the stones. The stones should be embedded into the soil/sand/cement to approximately two-thirds of their height, keeping the tops of the stones level to

one another. You can use the finer pea gravel as filler, but remember, you must place it vertically, not just throw it in place.

4. Use the two-by-four to help you keep the stones level as you go along by pressing them lightly as you complete a section.

5. Continue until your design is complete. Using more dry cement, dust the entire top of your stepping stone. This will bind the surface and fill in any gaps. Spray with water, and let settle. Dust again, and spray lightly. The dry cement will settle the stone and help to keep it in place.

Bibliography

Chester, Jonathan and Weinstein, Dave, *Berkeley Rocks: Building with Nature*, Ten Speed Press, California, 2007.

Long, Charles, *The Stonebuilder's Primer: A Step-By-Step Guide for Owner-Builders*, Firefly Books, New York, 1998.

McRaven, Charles, *Stone Primer,* Storey Publishing, Massachusetts, 2007.

McRaven, Charles, *Stonework: Techniques and Projects,* Storey Publishing, Massachusetts, 1997.

Snow, Dan and Mauss, Peter, *In the Company of Stone: The Art of the Stone Wall,* Artisan, New York, 2007.

Stanley, Tomm, *Stone House: A Guide to Self-Building With Slipforms,* Stonefield Publishing, New Zealand, 2003.

Author Bio

Brenda Flynn is a freelance writer and artist who lives in Ormond Beach, Florida. She has always had a passion for rocks and stones, and she studied with many artisans from the Dry Stone Conservancy in Kentucky, learning to rebuild dry stone walls from the Civil War era along the beautiful countryside of Kentucky, where she lived for six years. Flynn has worked in masonry and brick, building many outdoor projects, such as barbecues, patios, fences, and walls.

Index

E

F

G

H

I

Igneous, 19-20, 31-33, 36, 39, 5

L

M